RJ Toftness

UNLOCK THE MYSTERY TO MATH

Discover Why You Failed in Math

RJ Publishers

Hollywood, California

Unlock the Mystery to Math

Discover Why You Failed in Math by RJ Toftness

Published by
RJ Publishing
554 N. Mariposa Ave
Los Angeles, CA 90004, U.S.A.

Unattributed quotations are by RJ Toftness

Copyright © 2008 by RJ Toftness

Edition ISBN Soft cover 978-0-9821469-0-6

First edition 2008

Cover and illustrations by Michal Toftness

Unlock the Mystery to Math: Discover Why You Failed in Math by RJ Toftness

ISBN 978-0-9821469-0-3

Printed in the United States of America

CONTENTS

Introduction

How can I get a better grade in math? Why do some snarl at the idea of math? Why did national math scores start falling in the mid-1960s? If we could unlock these mysteries, then there would be hope.

I have been asked, "Who is this math book for?" The information in this book has been presented to middle school, high school, college students and older. The response has been, "Oh my gosh! I didn't know that before." It was an eye-opener for many students and adults.

I was curious to know how students *felt* about math. On a sunny spring afternoon, I visited a college campus in Los Angeles, California. The responses varied wildly to the question, "HOW DO YOU FEEL ABOUT MATH?"

"I love MATH," exclaimed a bright-eyed girl. She could not contain her enthusiasm when asked the question. Next to her was sitting an older polite gentleman stating, "Well, math is necessary. It's needed in many areas of life."

I hate MATH!

Moving through the campus, a student questioned let out a big yawn and said, "Whatever."

Another student nearby yelled, "I hate it!" Next to him a girl said, "It's the book's fault because it's so complicated. If it was easier to understand, I could at least get a **C** for a grade."

Continuing on, a young lady sitting alone on the grass said, "It's scary. I get so nervous each time I have to take an exam. That's why I don't do so well."

Close by, another girl responded, "I cried a lot because I just didn't get it. I tried to understand it. I would stay up late at night to study for a test but it just didn't help."

To others, math was just a hopeless, useless subject and they gave up.

We can probably relate to these students. Maybe at one time you had similar feelings about math. Whatever our feelings have been, there is hope for a better understanding of math.

While tutoring students, it became apparent to me there was a basic step missing. The material in this book was applied and as a result students' grades began to rise.

The intention of this book is to present the basics of math so some relief can be given to those who are stressed out by it. Even if a student is doing well in math, this information will possibly enlighten him/her even more.

Are You Brighter Than a Grade School/High School Student?

Math quizzes are available for each grade in chapter VIII of this book. Take the tests and find your strong and weak points in math. Test answers can be found on pages 86–91.

If fractions are found to be a weak area, then it is suggested to obtain the future publication, "Unlock the Mystery to Fractions." A series of books will go over specific areas of math.

Chapter I

The First Missing Link of Math

First off, *math* is short for *mathematics*. Math is taught in pre-school through college. Math has several branches.

 Math is made up of four branches:

> 1) *Arithmetic*
> 2) *Algebra*
> 3) *Geometry*
> 4) *Calculus*

These four branches make up what is called MATH.

Therefore, math is a very big subject that has four parts. A chart would give us a clearer understanding (see next page).

Knowing that math is made up of four parts gives us a place to start. From this fact, one can move forward to learning the branches of MATH.

This is the FIRST missing link to understanding math.

The Math Chart

As you can see, math has four branches or parts. Each one fits together like a puzzle. Arithmetic is first, then Algebra and next is Geometry. Some students go on to study Calculus. This is the order math is studied while going through school.

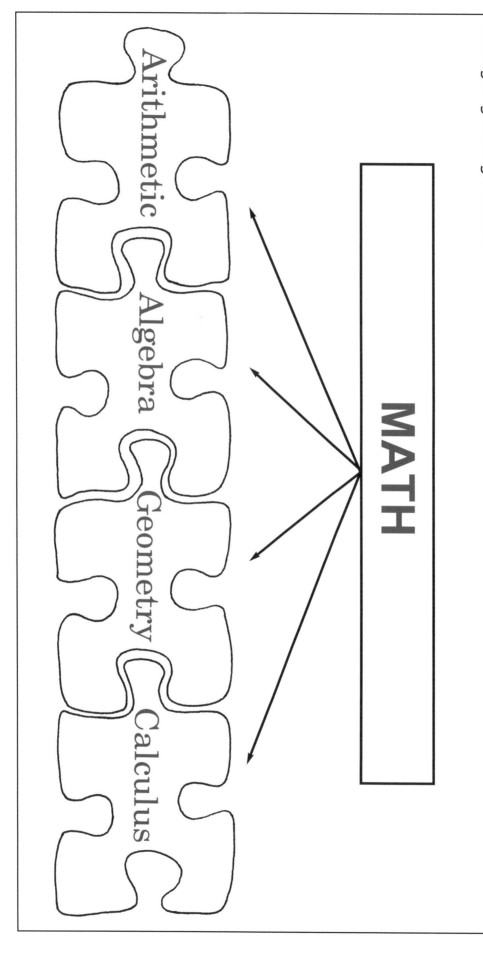

MATH

Arithmetic

Algebra

Geometry

Calculus

Lost

The smell of salt was around him. He was on deck looking outward, enjoying the setting sun. It was like a painting on someone's wall. Nathaniel had been taught at a young age to sail. Now at 22, he was out to sea on his own. He knew the tricks of sailing, but for his safety a GPS (Global Positioning System) was on board.

Nathaniel anchored for the night and went below to his cabin. He fell asleep quickly from the gentle rocking of the boat.

Time passed, then suddenly a loud thunderclap woke him up. The rain started pounding. The lightning lit up the cabin. The strong winds lashed against the boat tossing it back and forth. The motion was sickening. A hard jerk threw Nathaniel out of bed, slamming him into the table. All went black.

The next day, dazed and in pain, he slowly pulled himself up from the wet floor. Once on deck, his spirits fell. "No land in sight. The storm has pushed me out to sea."

Back in his cabin, desperation grew when he noticed the GPS was smashed on the floor. "At least I have my cell phone," thought Nathaniel. However, opening it, he found that water had damaged the phone.

"What will I do? I have no way to let anyone know where I am."

9

Nathaniel remembered reading stories about men being blown **off course** and eventually the boat was found with a dead sailor. "Is this my destiny?" thought Nathaniel. "The Pacific Ocean is so big. I have to think of something."

As night time was nearing, Nathaniel looked at the stars in the sky. He remembered reading about Christopher Columbus sailing the Atlantic.

"Columbus didn't have a GPS, let alone a cell phone. The stars can guide me back home." His spirits lifted. "I know the location of the Big Dipper and from there I can find the North Star. Then I can travel eastward and eventually hit land."

With a glimmer of hope he started his motor. "At least the engine still works." As he traveled throughout the night, daybreak slowly wiped away the stars. Off in the distance he saw something — a speck of land?

Nathaniel yelled, "YES, LAND!"

Going through math can sometimes feel like Nathaniel did, floating in the ocean, lost with no direction. But once he found the North Star, Nathaniel was able to steer his boat in a straight line back to land.

What if a student had something to use as a stable fact in math? He or she could then walk with some certainty and take the steps through math without feeling confused.

The first stable fact is math has four parts or branches. This gives us a starting point and from this we can build a good understanding of math.

Chapter II

The Second Missing Link

Teachers used to stress the importance of

☞ *Reading, Writing and Arithmetic.*

These are good facts when starting to learn.

1. Reading
2. Writing
3. Arithmetic

Hum....
What is
Arithmetic?

One thing, however, was constantly bumped into while tutoring students in math. The students did *not* know what the word *Arithmetic* meant.

I decided to survey students in the public schools of the Los Angeles area. This survey was to find what students knew about arithmetic.

One simple question was asked. "Do you know what arithmetic is?" To make sure the student knew what I was talking about, the word a*rithmetic* was written and shown to him/her.

It was found 92% of the students in grade school (mainly 4th and 5th graders) had *never* seen nor heard this word.

Did you ever
see this word?

Arithmetic

NO 90%
YES 8%

The same survey was asked of middle school students. The results were similar. Of those students, 90% did not know what arithmetic was.

I decided to ask college students if they knew what arithmetic was. It was found 40% did **not** know. The general response from the college students was, "Doesn't it have something to do with math?"

In all the schools visited, the total number surveyed was 293 students.

math

An active search was carried out to find if textbooks explained arithmetic. It was found that modern-day textbooks from 1st grade up did **not** explain the word *arithmetic*. The word was slowly removed from textbooks in the 1960s onward. Interestingly enough, national math scores started slipping in the mid-sixties.

So what happened? Did the math books throw away the word for a better word? Why was the word dropped out of use? Maybe by error, or the meaning of the word got lost.

It is not the intention of this book to unravel this mystery. The main point is that math has four branches and arithmetic is a vital subject that needs to be understood.

In chapter I, a chart showed the four subjects of math. The first section to be taken up is arithmetic.

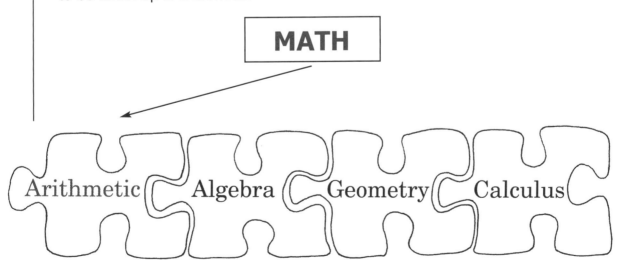

So, What Is Arithmetic?

Arithmetic is using numbers in

Adding

What will dinner **add** up to?

Subtracting

Darn, I have to **subtract** 10 points because of the penalty!

Multiplying

I can find the total by multiplying!

Dividing

Wow! How will I **divide** the profit?

Fractions

Percentages

Decimals

Counting

Problem Solving

Measuring

Example: Chuck has $12,000 in a savings account at 6% per year. How much will he have at the end of 1 year?

Answer: $12,720

We can see from the last two pages arithmetic is a basic part of math. A logical approach has to be taken or we have a student drowning in an ocean of data and facts.

When shown the important points of a subject, he or she has a better chance of improving in the subject.

Once a student has the important facts of a subject, then all the other data will fall into place. Textbooks omitting the precise differences in the branches of math could possibly be the first downfall for students.

This material has been presented to students who were tutored in the last few years. The results and successes speak for themselves.

The Arithmetic Chart

A chart has been drawn up to give a clearer picture of arithmetic. These 10 areas that make up arithmetic can be useful, as will be shown.

This chart will assist a student who is doing poorly in arithmetic. Also, a student starting arithmetic can be given this chart. It will provide a map of the journey through the subject. Even adults were surprised by the information.

All too often, a student runs for help when it's too late. The student stuck in advanced math is usually missing the basics. This can be very stressful for the student because he/she will have all of his or her attention on trying to get a passing grade on the current material. To go back to the basics can be tough, but the truth is IT IS what is needed for a true understanding. It will take some *burning of the midnight oil* but it will be well worth it.

Arithmetic

Adding	Subtracting	Multiplying	Dividing	Fractions	Percentages	Decimals	Counting	Measuring	Problem Solving
1,068 + 988 —— 2,056	3,007 − 1,995 —— 1,012	98 x 21 —— 2,058	$$\frac{94}{52\,)\overline{4{,}888}}$$ or 4888÷52=94	$\frac{7}{18}$	65%	.2387	1, 2, 3, 4 and so on	10.6 gallons	Monthly plan for a cell phone is $59.95 and $.10 for every minute over 1,000 minutes. If a person goes over by 10 minutes, how much will the monthly bill total? Answer: $60.95

Knowing well these 10 parts of arithmetic will take you into advanced math with certainty.

10 Parts of Arithmetic

Grade Yourself

This chart can be used to find your strong and weak areas in arithmetic. Place a grade of **A, B, C, D or F** on the lines below. For example, if fractions are a weak area, then mark a **C or D**. **Grade** yourself in each section.

① **Adding** _____

② **Subtracting** _____

③ **Multiplying** _____

④ **Dividing** _____

⑤ **Fractions** _____

(Adding, subtracting, multiplying and dividing fractions)

⑥ **Percentages** _____

(Finding a percent of a number, etc.)

⑦ **Decimals** _____

(Adding, subtracting, multiplying and dividing decimals)

⑧ **Counting** _____

(This deals with counting also time and money)

⑨ **Measuring** _____

(Measuring lengths using inches, feet, centimeters, meters. Measuring amounts, for example, ounces, pounds, liters, etc.)

⑩ **Problem solving** _____

(Example: Uri has 5 gallons of gas in his car. He wants to travel to Jerusalem which is 90 miles away. He gets 35 miles to a gallon. Does he have enough gas to do a round trip?)

Answer: No, he will run out of gas 5 miles before getting home.

Now that the exact area has been narrowed down, the difficulty can be approached without thinking everything has to be learned over again. It may be that a simple idea wasn't understood. So, grasping the idea and doing some problems to gain certainty is all that is needed. It doesn't have to be a long, difficult job.

It's been found some students struggle with fractions. The basic idea of fractions is what is missed. Once that is understood, the student can fly through with ease. Other areas that come up as difficulties are multiplying and dividing. It might be hard for an older student to accept this, but time spent on the basics will help in algebra and geometry.

There are many books and Internet sites that can assist a student once the exact area has been located on page 18. Future publications will cover these areas. Also, recommended books and Internet sites are listed on page 95. Take the quizzes in the back of this book. These tests will help find your strong and weak areas in math.

As a side note: arithmetic problems are usually done vertically (up and down), as shown in this example.

$$
\begin{array}{r}
359 \\
+\ 23 \\
\hline
382
\end{array}
$$

You will also see problems being done horizontally as shown here.

$$359 + 23 = 382$$

Arithmetic is the foundation for advanced math.

19

From F to A+

Self-discipline is vital when doing arithmetic or any math problem. In fact, it could mean the difference between an **"A"** or **"F"** on tests. Here's an example. While doing a problem on a test, a student might scribble on a scratch paper the numbers of a problem. But when this is done the smallest mistakes can happen. For example, adding numbers like this will result in the wrong answer.

761
+44
1,201

Wrong

Here is an actual example: A high school student doing a geometry problem needs to find the diameter of a circle. The only information given is the area of the circle which is 271.6 square inches. She works out the problem to where the radius is 9.3 inches. All she has to do is multiply the radius by 2 and that will give the diameter of the circle. When multiplying 2 x 9.3, she gets 19.6 inches. What was her error? That's right 2 x 9.3 = 18.6 inches. This mistake cost her because of rushing through the last part of the calculation. [Note: Diameter, radius and area are explained in the glossary.]

All too often, while tutoring students, it was found the student could not read the numbers he or she wrote down in the first place. So self-discipline is the key. Rushing through is a sure route to a lower grade. It is frustrating to find that a silly mistake was made in arithmetic when we actually knew how to solve the problem in the first place.

It goes without saying, good sleep, food, vitamins and exercise will help. So, calm down and focus is the way to approach this. Writing the numbers in straight columns will help you tremendously.

$$\begin{array}{r} 761 \\ +44 \\ \hline 805 \end{array}$$

Correct

Do not worry about your difficulties in math.
I assure you that mine are greater.

—Albert Einstein
Scientist

Our First Years of School

Arithmetic is first learned in our early years of education. Open any math (arithmetic) book. What do we see? Obviously we see

 1) Words

 2) Symbols (+, =, ÷, etc.)

 3) Numbers

Yes, the author will put in pictures and charts, but the ink on those pages is mainly forming words, symbols and numbers. In the next chapter we will investigate numbers.

✎ Check Your Knowledge 📖

1) The four branches of math are 1)_____
                                           2)_____
                                           3)_____
                                           4)_____

2) Which one is correct?
       A) Arithmetic is a branch of math.
       B) Arithmetic is advanced math.
       C) Arithmetic and math have no relationship.
       D) All of the above.

3) How would a stable fact or idea help a person in study? (For example: Nathaniel used the North Star to find his way home.)

4) Name the 10 parts of arithmetic.

    1)_____   6)_____
    2)_____   7)_____
    3)_____   8)_____
    4)_____   9)_____
    5)_____   10)_____

5) What can happen if one is sloppy while doing math problems?

6) True or False? The word *arithmetic* is known by all students.

7) What is the best way to get a good grade?
   A) Study all night for the test.
   B) Rush through the problems and hope they are right.
   C) Calm down, focus and write the problems in straight columns.
   D) None of the above.

8) What are the three things you will find in any math book?
   1) _____
   2) _____
   3) _____

9) Arithmetic is (circle the correct answer)
   1) a branch of math;
   2) the skill of using numbers in adding, subtracting, etc.;
   3) a basic part of math used in advanced math;
   4) all of the above.

Answers:
1) Arithmetic, algebra, geometry and calculus.
2) A.
3) When a student has a stable fact he/she can then align everything else; therefore a student isn't lost in a lot of data.
4) Adding, subtracting, multiplying, dividing, fractions, counting, percentages, decimals, measuring, problem solving.
5) Lower grade.
6) False.
7) C.
8) Words, symbols and numbers.
9) 4.

Arithmetic is being able to count to twenty without taking off your shoes.

—Mickey Mouse

Chapter III

Are Numbers Haunting Us?

Numbers surround us. In fact, we are walking with numbers. They are on our clothes (size), shoes (size), underwear (size), watch, coins in our pockets, even on our glasses and possibly on the inside band of a ring. What about the wallet? Money, credit cards and library cards all have numbers. How about our cell phones? How many phone numbers do you have?

In our modern world we can't get by without numbers. What time did we wake up this morning? Did we have to get to work or school? How long did it take? What road or freeway number did we take? Was there a bus to catch? What bus number? Did we drive a car? How fast? Did we turn on the radio? If so, to what station number? What number is our home, apartment? Catch a favorite TV show? What channel and at what time? Microwave some food? How many minutes was it set for?

This is just a glimpse of the numbers we use. In fact, during any 15-minute period of our waking day we use numbers. You say, "Not possible." What page are we on right now?

Since we can't avoid numbers, let's find out more about them.

Roman Numerals

☞ 27 BC

(Roman Empire begins)

Roman numerals were used by the people of the Roman Empire.

I = one
V = five
X = ten
L = fifty
C = one hundred
D = five hundred
M = one thousand

Note: Roman numerals did not include zero.

☞ AD 476

(Fall of the Roman Empire)

Roman numerals continued to be used for another 600 years after the collapse of the Roman Empire.

☞ AD 1000

Around AD 1000 Roman numerals were thrown away. Why?

Maybe Roman numerals were just too hard to work with. Try dividing Roman numerals and you will see why. Example: **XII ÷ V = ?**

OUR NUMBERS

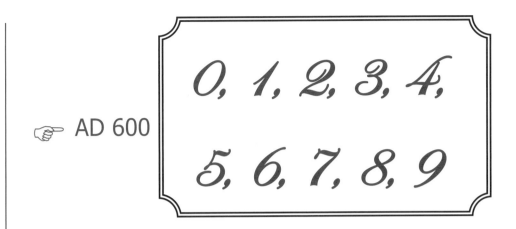

☞ AD 600

Another form of numbers was developed in India around AD 600. They were easy to use and based on 10 (0, 1, 2, 3, 4, 5, 6, 7, 8, 9). They were maybe thought up because we all have 10 fingers.

During that time, people of India traded with their neighbors in Arabia. The Arabians liked the numbers so much, they used them in their everyday life.

☞ AD 1000

I'LL TRADE MY SHEEP FOR THOSE NUMBERS?

27

☞ AD 1100

Around AD 1100, Arabians from northern Africa migrated to Spain and brought the number system with them. They also built Spain stronger than any other country in Europe at that time.

☞ AD 1200

About AD 1200, an Italian man named Leonardo Fibonacci wrote a book that introduced the Arabic number system (0, 1, 2, 3, 4, 5, 6, 7, 8, 9). These numbers caught on fast in Europe since they were much easier to use than the awkward Roman numerals.

As a result, these numbers became internationally known and are used throughout the world today.

Numbers came from India, traveled to Arabia then to Europe. From Europe the numbers spread to the rest of the world.

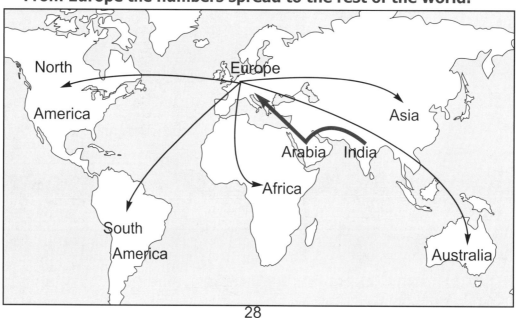

NUMBERS HAVE NAMES

The Romans called their numbers the *Roman numerals*. Our numbers also have a name. They are called the *Arabic numerals*. In truth, they should be called the *Hindu numerals* (Hindu referring to India), since these numbers 0, 1, 2, 3, 4, 5, 6, 7, 8, 9 started in India back in AD 600. Once in a while we will see these numbers referred to as *Hindu-Arabic numerals.*

Roman numerals   ☞ I, II, III, IV, V, VI, VII, VIII, IX, X

Arabic numerals   ☞ 0, 1, 2, 3, 4, 5, 6, 7, 8, 9

Cardinal numbers ☞ One, two, three and so on

Ordinal numbers   ☞ First, second, third and so on.

So, out went the Roman numerals and in came the Arabic numerals we use today. We still see the Roman numerals. For example, Roman numerals are on clocks and in various other things like the chapter numbers of this book. However, they are not as practical as the numbers we use today.

✎ Check Your Knowledge 📖

1) Write the Roman numerals.

2) Why were the Roman numerals disregarded eventually?
   A) Too awkward to use B) Arabia introduced Arabic numerals C) A new number system came along. D) All of these are true

3) Where did our numbers originally come from?
   A) India   B) Roman Empire   C) England   D) New York

4) What is the name of our numbers?
   A) Roman numerals   B) Arabic numerals   C) Numbers   D) American

5) Could you get through one day without using numbers?

6) What date did our numbers come into existence?
   A) AD 600   B) AD 1100   C) AD 1976   D) 27 BC

7) True or False? Roman numerals became useless because of the collapse of the Roman Empire.

Answers: 1) See page 26 2) D 3) A 4) B 5) Answer varies 6) A 7) False

---

Brain Teaser

One sheet of paper is about 1/1,000 inch thick. You cut the paper in half and placed the pieces on top of each other. Then cut that in half and place them together.
   If you did that 50 times, how tall would the pile of paper be?

Answer:
About 17,000,000 miles

Brain Teaser

If Julius Caesar put 1 cent in the bank 2,000 years ago and the interest was compounded each year, how much would he have in the year 2000?

Answer:
45 zeros following $150 (Lot of $)

The things of this world cannot be made without a knowledge of mathematics.

— Roger Bacon
English scientist (1214–94)

♘ One Horse, of Course ♘

Mr. Tom Hippityhop's death was sudden and a surprise to the people of Happy, Texas He was a wealthy man and a proud owner of 17 sleek, beautiful horses.

Throughout the community, Mr. Tom was known as a smart and sometimes mischievous person.

Mr. Hippityhop's wife slowly recovered from the shock. Weeks later, while going through his papers, she came across a will written by him.

To my Dear Wife, Son and Daughter,
I know these are difficult days for you. I have one wish I would like to be carried out after I pass away. As you know, the 17 horses are my pride and joy. I want to give these horses to you so that they will be well taken care of. Therefore:

you, my darling wife, get one-half of the horses;
you, my son, get one-third of the horses;
and my beautiful daughter, you get one-ninth of the horses.
The executor of this will shall be Mr. Mason, our town teacher.

Tom Hippityhop
Happy, Texas
April 3, 1898

Mrs. Hippityhop was relieved to turn this problem over to the town teacher.

"Hum," said Mr. Mason as he stroked his beard. "This could be a real problem dividing the horses evenly amongst all of you." Mr. Mason thought to himself, "I can't divide these without having to split one horse into sections." After a moment, Mr. Mason said, "Within a week, I will be back with an answer."

Mr. Mason couldn't sleep that night. He kept thinking, "If only Mr. Tom had 18 horses, this problem would be easy to solve." All of a sudden a light bulb turned on in his head. He knew how to solve this problem.

The following week, Mr. Mason rode his horse to Hippityhop's farm.

"Good day, Mrs. Hippityhop and children," said Mr. Mason as he got off his horse. "I have the answer to the problem. However, before we start, I am going to loan my horse to you for a few minutes."

"Mrs. Hippityhop," said Mr. Mason, "You get 9 horses, which is one-half of all the horses. Young man, you get 6 horses which is one-third. And young lady, you get 2 horses which is one-ninth. As you see, 9+6+2=17."

With a sigh of relief, Mrs. Hippityhop said, "Oh, thank you, Mr. Mason!"

Mr. Mason mounted his horse, tipped his hat and with a smile said, "Have a pleasant day, folks."

Only the educated are free

— Epictetus (Greek philosopher)
AD 55–135

Chapter IV

Oh No! Not Algebra!

The next branch to take up is algebra.

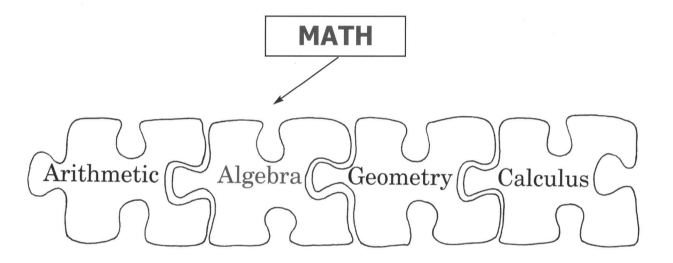

Algebra came from Arabia. Not only did our numbers come from Arabia, but we got our second branch of math (algebra) from there.

A wise man named Al-Khwarizmi is considered the father of algebra. He lived in Baghdad (Iraq) around AD 850.

The word a*lgebra* means "reunion or completion." It originally meant *bringing together a broken bone.*

In a way, we could say we are bringing equations to a *completion* by finding the answer.

The beauty of algebra is that more complicated problems can be solved which can't be solved with simple arithmetic. This chapter covers the basics of algebra. A future publication, *Unlock the Mystery to Algebra,* will go into more detail.

☞ The Guessing Game

Algebra books love to use the end letters of the alphabet. Therefore, we see **x , y** used quite often. All we need to do is find the *correct* number to replace the letter. The challenge for us is to find what number will answer the problem correctly. So letters are used as shown here.

X + 9 = 12   or   **Y** + 9 = 12

In truth, any letter could be used.

T + 9 = 12   or   **A** + 9 = 12

But it all boils down to one thing:

What? + 9 = 12

The symbols of arithmetic (+, −, ÷, x) are also used in algebra. Algebra introduces other symbols, for example:

Parentheses ( )   Braces { }   Brackets [ ]

Using these symbols would look something like this. Solve

$$[5 + \{- 7 + 19 + (9 \div 3) - 2\}]$$

If you got 18 as an answer, good job. ☝ If not, do the parentheses first $(9 \div 3) = 3$, then the braces $\{- 7 + 19 + 3 - 2\} = 13$, and finally do the brackets $[5 + 13] = 18$.

The rule is

      1) Calculate what is inside the parentheses first.   ( )

      2) Then work on what is inside the braces.   { }

      3) Finally, do what is inside the brackets.   [ ]

☞ Negative Numbers

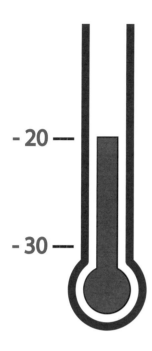

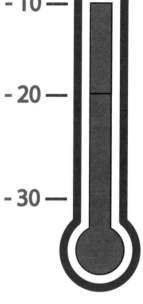

Students sometimes have difficulty with negative numbers. Whenever possible, algebra problems should be matched with real-life experiences, then the world of math blooms in front of the student.

Let's take a problem like this:
$$-20 + 10 = ?$$
Applying this to temperature will make sense. Let's say it is 20 degrees below zero. (Southern California people may not experience such a temperature but the folks in Wisconsin can tell you some cold winter stories.) Now imagine getting 10° warmer, so the temperature is a "balmy" 10 degrees below zero ($-10°$). Going back to the problem, $-20 + 10 = -10$ will now make more sense.

☞ The Balancing Act

Another symbol that is used frequently in algebra is the equal sign (=). The equal sign tells a person that one side has to be the same as the other side of the (=) equal sign. The equal sign is like a balance— both sides being the same.

If we had an equation like this: $x + 25 = 35$ and replaced x with 11, the balance would **not** be even, since we know 36 does **not** equal 35. We know the answer for x is 10; then it would balance.

Understanding the symbols of algebra is the first step to mastering the subject. For example, in algebra multiplying can be written in four different ways, as shown here.

$$3 \times 3 = 9 \qquad 3(3) = 9 \qquad 3 \bullet 3 = 9$$

Or nothing need be shown when multiplying 3 and Y, as shown here.

$$3Y = 9$$

Algebra builds on itself like a pyramid. Understanding the basic blocks will help a student figure out complicated equations. So what are the basics of algebra?

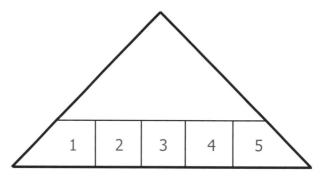

1. Knowing the words, starting with the word *algebra* (page 33).
2. Understanding what an equation is (page 38).
3. Having arithmetic down cold (10 parts, page 18).
4. Learning the symbols used in algebra (pages 34, 35).
5. Solving an equation (see next section).

☞Solving an Equation

Every equation has an equal sign (=), as shown on page 35. To solve a simple equation in algebra, it can be done in your head. For example, a problem like this:

$$X - 23 = 10$$

We could say, "*What number, if you subtract 23, would give us 10?*" Of course, **X** would be **33**.

However, in pre-algebra, it's suggested to do the following method. It will help on future complicated algebra problems. The goal is we want "**X**" all alone. Let's use a seesaw for the sake of simplicity.

The way to get the "**X**" by itself is to add 23 on the left side of the equal sign. The rule in algebra is

Whatever is done on one side of the equal sign is done exactly on the other side.

Remember, we have to keep the seesaw balanced.

So we add 23 on both sides of the equal sign.

$$X - 23 + 23 = 10 + 23$$

Adding 23 to the left cancels out − 23, because − 23 + 23 = 0. Now we have "**X**" by itself. Therefore

$$X = 33$$

Problems can be checked to see if the answer is correct by replacing X with the number found.

$$33 - 23 = 10 \quad \text{Very good, it's right!}$$

☞ Parts of Algebra

Knowing the meanings of words of any subject is the key to understanding it. Algebra has many terms. The words used most often are

1) Equation   2) Coefficient   3) Expression   4) Exponent   5) Variable

The best way to explain these words is to show them in action.

$2x + 5 = 15$ This is an equation (all equations have an equal sign).

$2x + 5 = 15$ The coefficient is 2. It is the number or letter that is the multiplier. Another example would be this: 7y where 7 is the coefficient.

$16x$ This is an expression. Note, an expression does not have an equal sign but stands alone.

$9^2 = 81$ 2 is the exponent. It shows how many times the number is multiplied by itself (9 x 9).

$2x + 5 = 15$ Variables in algebra are letters standing for a number. For example, in this equation $2x + 5 = 15$ the x is the variable.

It is wise to get these down so you know them with a snap of the fingers.

There are many more words in algebra; however, this is a start to getting a better grade in the subject.

Algebra could be divided into two sections as shown on the graph below.

Algebra

Simplifying expressions	Solving equations
$\dfrac{2y}{2y}$ This expression can be simplified to 1.	$7x + 30 = 170$ $x = 20$

☞ Grade Yourself

A list of different algebra calculations can help find the weak and strong areas. Grade yourself with A , B, C, D or F. (Note: This is not a full list but covers some of the basic areas of algebra.)

①   Solving Simple Equations.   _____

         Example: $2X + 6 = 10$

②   Simplifying to Simplest Form.   _____

         Example: $\dfrac{36x^2}{12x} = 3x$

③   Exponents.   _____

         Example: $10^2 = 100$

④   Factoring.   _____

         Example: $2x + 6 = 2 (x + 3)$

⑤   Negative Numbers.

         Example: $-5 - 3 = -8$

⑥   Graphing Straight-Line Equations.   _____

         Example: $y = 3x + 6$

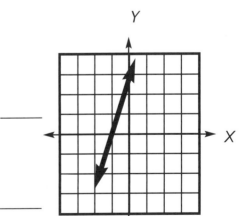

⑦   Inequalities.   _____

         Example: $4x + 4 > 20$

                     $x > 4$

As we can see, having arithmetic down cold will make algebra easier.

 Check Your Knowledge 📖

1) Where did the word *algebra* come from?

    A) Africa   B) Arabia   C) England   D) Russia

2) What does "*reunion or completion*" have to do with algebra equations?

3) What does this sign $=$ mean?

    A) Subtract twice   B) Equal   C) Equation   D) Variable

4) What are five areas you should know to master algebra?

    1)_____   4)_____

    2)_____   5)_____

    3)_____

5) Solve mentally $34 - 2Y = 8$

Answers: 1) B 2) See page 33 3) See page 36 4) $Y = 13$

Brain teaser

You are offered two different ways of getting paid by a boss for a part-time job. Which salary would you take?
Starting pay of $8,000 a year with $400 increase every year or $4,000 every six months with $100 increase every six months.

Answer: You will make more with the second plan.

Brain teaser

A cell-phone company offers two payment plans. One plan is a flat rate of $195.95 each month. The other is $49.95 for 500 minutes, and over 500 minutes it's 10 cents for the first 5 minutes, 20 cents for 10 min., 40 cents for 15 min., 80 cents for 20 and so on. You always use 600 minutes each month. Which plan is better?

Answer: $195.95 is better. The other plan will cost $254.75/month.

The Rope Stretcher

*S*weat rolled down his forehead to the hot sand. Nasu was used to the heat, but as he gazed at the pyramids he wondered how much heat he could withstand. Luckily for him, he had grownup along the Nile River, which was somewhat cooler because of the breeze off the water.

"Nasu," cried his mother, "get back to work or else the boss will pay us less." Nasu shouted back, "I hate this work!"

From an early age, Nasu had worked long hours in the fields along with his mother and father. Nasu owned very little. His most treasured item was a rope his grandfather had left behind. He would make different shapes with the rope every chance he had.

Nasu could still hear his grandpa say, "There is a special shape you can form with this rope, Nasu. Tie knots equal distances apart. Make a triangle so one side has 3 spaces, the next side has 4 spaces and the diagonal side has 5 spaces. This triangle is special because it's a right triangle. One side looks like a palm tree growing straight out of the ground."

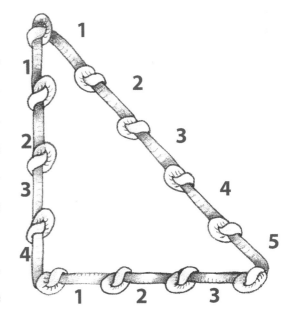

Nasu thought his grandpa was a genius, because it worked each time as long as he tied the knots equal distances apart.

Each spring the Nile River flooded. The rushing waters would wash away the markers dividing the farmers' land.

41

One day, Nasu witnessed an argument between two farmers.

"You steal my land after the floods, you thief!" yelled one farmer.

The large man looked down at the other and said, "I will take whatever land I want and YOU will not stop me!"

Farm owners would die mysteriously due to the conflicts.

King Ramses II of Egypt posted a reward stating, *Whoever solves the boundary problem between the farm owners shall be rewarded. Give your solution to the king's court on the twelfth day.*

Ramses II

On that day, after several people presented their solutions, the king motioned to the young boy dressed in rags to step forward. Nasu pulled the rope from his bag.

"Sir," said Nasu. "I can solve the problem with this rope." The king's court laughed. The king quickly silenced them.

"Continue," said the king. Nasu showed his special triangle. He explained how square plots of land could be made by using the right triangle.

After viewing all the solutions, the king stated, "A decision will be posted tomorrow."

The following day news spread quickly. Nasu raced to the village. An elderly man read the king's decision to Nasu.

A young boy named Nasu gave the best solution. As of this day, land will be measured by Nasu's plan. Nasu is awarded a plot of land.

Ramses II

Nasu ran home to tell his parents the good news.

As the years passed, Nasu became very well known throughout the Nile valley as THE ROPE STRETCHER.

42

Chapter V

Geometry

Geometry is one of the oldest subjects, dating back to 3000 BC. It was used by the Egyptians for construction and measuring. Even though it is a vital subject, it relies on the use of arithmetic and algebra. Geometry takes the third position on the chart.

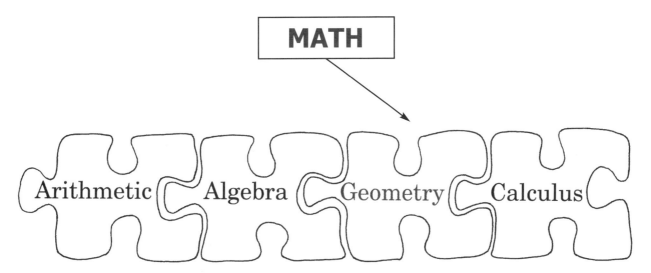

Examining the word *geometry*, we find *geo-* comes from the Greek word meaning *earth* and *-metry* means measure. We see *geo-* in words like

Geography: study of the earth's climate, animals, humans, etc.
Geology: study of the earth's rocks, crust, etc.

Therefore, *geometry* means "the measure of Earth."

Indeed, that is exactly what we are doing in geometry. Look around us. Everything that is made on Earth is measured. For example, this book was measured before going to print. Benjamin Franklin was spot on when he said, "No mechanical invention [can be made] without geometry."

Even though the Egyptians used geometry in measuring and building, it wasn't until 300 BC that a Greek man named Euclid wrote about it in a book called *Elements*. Euclid's findings and written information are still used today in our geometry books.

Space

We could describe geometry with one phrase:

HAVING TO DO WITH SPACE

Once we have an understanding of this, we are on our way to being able to use geometry in everyday practical applications.

We are not talking about *outer space,* although it could apply there also. When we say space, we are talking about distance between two points, area of a surface, or the space in a container. In fact, geometry could be broken down into four parts: a point (dot), a line, a plane (surface), and volume (solid figure). Showing these on a chart will make it easier to understand.

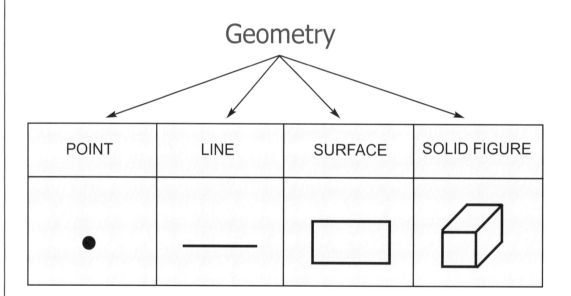

To have a point, a line, a plane (surface) or a solid figure we need SPACE to put it in. Without space, we would not have a subject called geometry.

The first three sections of the chart (point, line and surface) make up the subject called plane geometry. This can be shown as shapes like these:

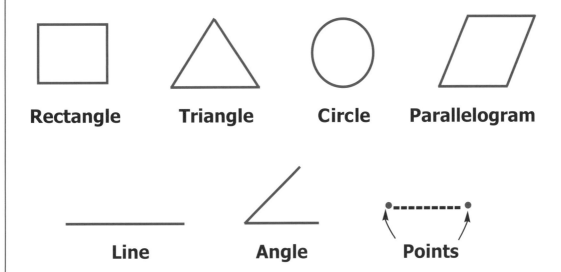

Rectangle   **Triangle**   **Circle**   **Parallelogram**

Line   **Angle**   **Points**

The last section (solid figures or volumes) makes up the subject called solid geometry. These are shown as

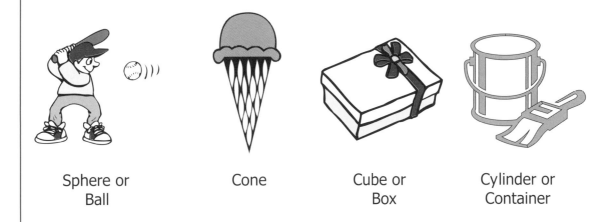

Sphere or   Cone   Cube or   Cylinder or
Ball   Box   Container

There are other solid figures in geometry. For example, pyramids, prisms, etc. Now having these two parts of geometry (plane and solid geometry), one can advance into the more difficult problems.

☞ Geometry in Life

Geometry is used in different areas of life. Carpenters building a home use it to have square walls and floors. See page 61 for more uses of math in everyday life.

How many yards will she need to sew for her living-room curtains?

Also, geometry is used for measuring angles and distances. An example would be to measure the land for a new building.

☞Grade Yourself

Geometry problems can be broken down into different sections. These are some of the parts of elementary geometry. Grade yourself in each section below with **A , B, C, D or F**.

① **Finding the area and perimeter of a shape.** _____

Example: A square is 5 feet on each side. What is the perimeter and area? *Answer: Perimeter is 20 feet. Area is 25 square feet.*

② **Finding the volume of a solid figure.** _____

Example: Find the volume of a tank which is 10 feet tall and has a diameter of 20 feet. *Answer: 3,140 cubic feet*

③ **Knowing shapes**. _____

Example: What is the name of a 5-sided figure? *Answer: Pentagon*

④ **Finding the length of the diagonal side of a right triangle.**

_____

Example: Two sides of a right triangle are 15 and 20 inches. How long is the diagonal? *Answer: 25 inches*

⑤ **Finding sizes of angles.** _____

Example: A triangle has two angles of 25 and 64 degrees. How big is the third angle of the triangle? *Answer: 91 degrees*

⑥ **Graphing** _____

Example: Graph the following points and connect the dots.
(2, 3), (−2, −1) and (2, −1)

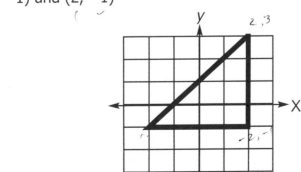

Note: These are just a few areas of elementary geometry.

 Check Your Knowledge

1) What *ONE* word could sum up the subject of geometry?

  a) Money   b) Space   c) Time   d) Energy

2) How does the word *space* relate to geometry?

3) What does the word *geometry* mean?
  a) Study of earth   c) Measuring Earth
  b) Study of rocks   d) Angles

4) What are the four parts of geometry?

  a)_____   b)_____

  c)_____   d)_____

5) What is the difference between solid and plane geometry?

Answers: 1) b 2) See page 44 3) c 4) See page 44 5) See page 45

Brain Teaser	Brain Teaser
A pipe is placed around the earth's equator, which is approx 25,000 miles. The pipe was made 20 feet too long. Rather than cut off the extra 20 feet, how far off the earth's surface would the pipe need to be raised to compensate for the extra 20 feet?	Where on Earth are you? You walk one mile south. Then you walk one mile east and next go one north. You find yourself exactly where you started. Where are you when you started the hike?
Answer: 3.18 feet off the ground.	Answer: North Pole

A TRIP TO P3S

The spaceship glided through darkness without a sound. On the lower right-hand corner of the monitor screen flashed the date and location [Year 3025 – Sirius]. It was an easy run taking only two hours between the two star systems, Vega and Sirius.

"Bring me the case with the confidential papers," commanded Captain RL to her junior. "We'll be at the landing pad within 10 minutes."

RL had commanded the craft for several years. Her job was to transfer confidential documents between the presidents of the star systems.

Approaching Sirius

10 minutes   Year 3025–Sirius

She was nicknamed **R**ocket **L**ady because of her ability to fly spaceships so well.

On the monitor screen flashed an order.

Captain RL: Report to the President's chamber upon arrival.

"This is odd," she thought. "I usually just drop off the documents to security and fly to my next destination."

As she walked into the president's chamber a bold voice filled the room. "Have a seat," said the president. "Captain RL, you are to go to P3S to receive an award."

"An award?" RL responded.

"Yes!" replied the president. "You are being commended for exceptional service to the Galactic Empire. Since P3S is a 31-day flight, you can supply your ship here with extra fuel and food."

49

On day 16 of the trip, the monitor flashed
TAC NONFUNCTIONAL
"What happened?" thought RL. The monitor screen flashed an answer.
<u>T</u>hought <u>A</u>ctivated <u>C</u>omputer not working

The Tech Support Team reported, "Sir, a new hard drive is needed for the computer. We'll have to return to the home base for a new one."

"NO!" said the captain. "We need a solution right now. We will not miss the ceremony on P3S."

"Horn goggles!" responded a junior.

"What are you talking about?" said the captain sternly.

"Horngoggles can help," replied the junior. "We call him that because of his thick glasses. He's always reading ancient books."

"Get him here!" snapped the captain.

Once Horngoggles was in front of the captain she said, "We need scientific calculations to manually fly the ship to P3S. TAC has crashed."

"Well," said Horngoggles, "I have read some ancient books about calculus." It's an old complicated math that has to do with force, direction, time and matter. I could figure out a formula in a day or so."

"You have two hours," ordered the captain. "Report back to me with the answer."

Two hours later, Horngoggles presented the solution to the captain.

RL commanded her juniors. "Use this to guide the ship, taking eight-hour shifts." RL turned to Horngoggles. "And you will sit to the right of me at the ceremony. Horngoggles responded, "Yes, sir!"

During the ceremony on P3S, RL turned to Horngoggles. "What is your real name?" Horngoggles responded, "Nerdoff." With a grin, the captain said, "I like Horngoggles better."

"Ya, me too," he responded.

"Sir?" asked Horngoggles. "Why is this planet called P3S?"

With a startled look, the captain replied, "I'm surprised you haven't run across it while studying the old books. It's Planet number 3 from the Sun. The ancient people used to call this planet EARTH."

Chapter VI

The Unmentionable

Column four completes the branches of math.

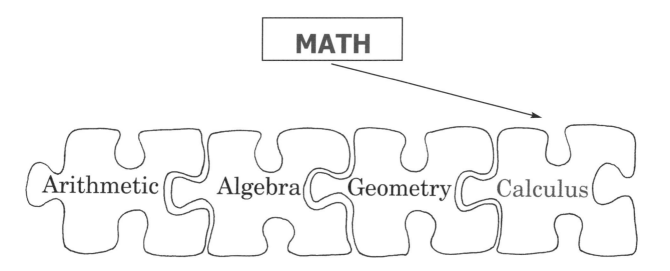

The word *calculus* comes from Latin meaning "pebbles" (stones). The early Romans used small stones for counting or "calculating" while doing arithmetic problems. However, as math continued to advance, especially during the 1600s, the word *calculus* took on a totally different meaning.

Calculus, simply put, is math which combines arithmetic, algebra and geometry to solve complicated problems.

Isaac Newton was a major contributor to calculus. Oddly enough, at the same time another man (Gottfried Leinniz) developed calculus also. A great controversy arose as to who developed it first. It was finally sorted out that both men contributed to it independently.

The development of calculus has advanced our civilization and solved problems unthinkable 200 years ago. From building tall skyscrapers to sending spaceships to the moon, calculus has been the backbone of our modern advancement.

Calculus could be divided into two sections, one dealing with rates of change and the other having to do with changing quantities.

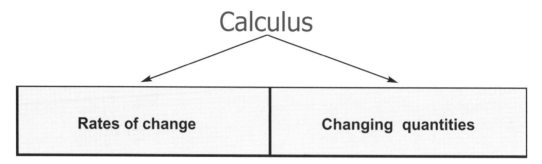

Calculus

Rates of change	Changing quantities

Examples of these two sections would be the following:

An airplane travelling 600 miles per hour would be an easy calculation if we wanted to know how far it travelled in 3 hours. (Multiply 3 X 600 = 1,800 miles.)

But suppose the plane doesn't fly at a constant speed because of changes in wind, air pressure and other conditions. Therefore a higher math is needed to calculate speed, distance and time. This higher math (calculus) can be very useful, especially when calculating the amount of fuel used in a flight.

Another example would be a cannonball shot from a cannon. By using calculus, one could determine where the ball would land, how far it would go and the time it took, also how fast the ball was going at any point.

Calculus is used in computer science, engineering, business, medicine, and by NASA, the military and in many other professional fields.

☞ Rates of Change

Rates of change could be shown as how far an object falls through space in a given time.

Imagine a scientist is in a hot-air balloon. He drops a marble from the basket. (This is not advised.)

Let's assume the scientist has an instrument to measure the distance. In 1 second the marble falls 7 feet. In 2 seconds it falls 28 feet and so on.

From these measurements he can work out a formula. The distance the marble falls equals 7 multiplied by time squared. (Distance = $7T^2$)

2 seconds

28 feet ●

Now we can find the distance at any time during its drop. Let's say the marble was dropped 10 seconds ago. Plugging the numbers into the formula, we see 7 x 10 x 10 = 700 feet the marble has fallen in 10 seconds. This branch of calculus is called by a fancy name—*differential calculus*.

☞ Changing Quantities

From the example above, if he wanted to find how fast the marble is falling at a particular place, the scientist could work out a formula. The speed of the marble is equal to 14 multiplied by the time. Another way to write this would be speed = 14 x time.

If we wanted to know how fast the marble is traveling at 10 seconds after it left the balloon, just plug in the numbers to the formula. So the speed the marble is traveling is 14 x 10 = 140 feet per second.

This branch of calculus also has a unique name—*integral calculus*.

Finding the quantity or the amount of something is also used extensively in different areas of life. To find the area of a solar panel, for example, all we need to do is multiply the width and length of the panel. We would have 60 square feet of solar panel.

This type of calculation is learned in geometry because we're dealing with a specific shape, as shown on page 45.

A person might think that this is the most used math applied in everyday living.

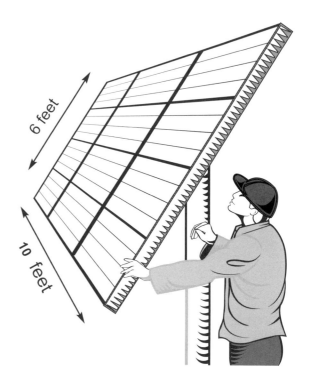

However, what if we wanted to find the number of gallons this swimming pool contains? It is not as easy as multiplying the length, width and depth. Since this swimming pool has curves, we'll need a more advanced math to find out the answer.

This is where calculus saves the day.

The discovery of calculus is very important. In fact, many scientists would say it is the most important branch of all math. But when it boils down to calculating the answer, in the end it still relies on good old arithmetic.

Isaac

Isaac, pay attention!" snapped the teacher's voice. The young boy looked startled. His classmates started to laugh but shut up quickly. Mrs. Carnapple gave a look that would give anyone the shivers, let alone make one silent.

She approached Isaac with a stern look. She grabbed the book he had on his lap. "Where did you get this?" she demanded.

"My uncle's home," Isaac responded quickly. Isaac knew better than to talk back. The last time he did, he was whipped several times and could not sit for days. The teacher threw the book on the table, startling the other students. "Do not bring this back to school!" she commanded.

Isaac had been born December 25, 1642, in a small farmhouse on the rolling land near Wool-shorp, England. From the start, he had done poorly in school. All Isaac really wanted to do was make gadgets and read.

At the age of 14, Isaac had to drop out of school to help his mother on the farm. Every chance he had he would be reading or working on the latest project.

After three years, his mother decided to send him back to finish high school and continue with college.

His enthusiasm to learn never diminished. On a warm August day, taking a break from his study, Isaac sat under an apple tree. What happened next changed the whole course of math and physics. An apple hit him square on his head.

From this insignificant event, Isaac observed an obvious action that had been viewed by prior generations. However, Isaac questioned, "Why should such a thing happen?" There was no hand pulling the apple down. From this curiosity, Isaac went on to discover the Laws of Gravity.

Years later, in 1705, standing before Queen Ann of England, he was the first scientist to be knighted: Sir Isaac Newton.

This story emphasizes that simple observations can lead to a greater understanding. Knowing the basics of math can give a student more certainty. Then one can advance into higher math without being overwhelmed.

> *Intelligence plus character—*
> *that is the true goal of education.*
>
> —Dr. Martin Luther King Jr.

Chapter VII

So, Does Math Add Up?

Is it possible to explain all the branches of math in one statement? We have covered the four branches and given a basic explanation of each branch. Math, in general, could be explained as

using numbers and letters to solve problems dealing with

★ *1. Time*
★ *2. Space*
★ *3. Energy*
★ *4. Amounts*
★ *5. Distances*

An example would be taking a trip in the car. If we want an idea of how much gas is needed, the total cost and how long it will take, we would need to solve the problem using amount, distance and time, which has to do with energy and space.

Without math our world would be a very complicated place to live in. Could we live without math? Well, we could but it would be tough. So, does math add up? I'll let you answer that one.

Which saves more energy?

I wish I had more **time** to sleep.

We have looked at the basics of math. These four branches are the ones most commonly heard in school. However, there are many other branches of math (hence the puzzle can be added to). For example, probability deals with chances and outcomes. Another branch is trigonometry, which has to do with angles.

Of the four (arithmetic, algebra, geometry and calculus) which is the most important? Yes, arithmetic is king; but if we were to talk to a scientist, he would probably say calculus is senior. But to arrive at the answer in calculus, we still need to use arithmetic.

Having a firm understanding of the basics, we can now advance into math with more confidence. It may be that not having this knowledge made math boring, disliked or hated, and resulted in tears shed or failure.

Building a house on shaky ground will result in a collapsed structure. However, building it on solid ground will give the house support for generations. With these basics, you now can climb to the stars.

Our greatest natural resource is the minds of our children.
— Walt Disney

MATH

Arithmetic	Algebra	Geometry	Calculus
1) Adding	1) Simplifing expressions	1) Plane geometry	1) Changing Quantities
2) Subtracting	2) Solving Equations	2) Solid geometry	2) Rates of change
3) Multiplying			
4) Dividing			
5) Fractions			
6) Percentages			
7) Decimals			
8) Counting			
9) Measuring			
10) Problem solving			

★Great Mathematicians★

500 BC ★ Pythagoras: Greek philosopher, contributed to the geometry we use today.

400 BC ★ Plato's teacher Socrates and his student Aristotle gave much of the philosophic reasoning to the Western world.

300 BC ★ Euclid: father of geometry.

0 ★ Roman Empire gave us the Roman numerals: *I, II, III, IV, V , VI, VII, VIII, IX , X.*

AD 400 ★ Hypatia: the first woman to have the greatest impact on math. Later, her writings were expanded by Descartes, Newton and others.

AD 600 ★ Indian culture gave Arabic society the numbers *0, 1, 2, 3, 4, 5, 6, 7, 8, 9.*

AD 1200 ★ Leonardo Fibonacci introduced the Arabic numbers to Europe.

AD 1500 ★ Blaise Pascal built a mechanical calculating machine.

AD 1600 ★ Descartes contributed to algebra and geometry.

★ Sir Isaac Newton: father of calculus and the Laws of Gravity.

AD 1700 ★ Carl Gauss contributed to advanced algebra.

★ Charles Baggage developed a number system used in computers.

★ Leonhard Euler: although totally blind, he invented many of the symbols which are used in advanced math (calculus).

AD 1900 ★ Albert Einstein advanced the ideas of energy and mass— $E=MC^2$ (*E* is energy, *M* is mass and *C* is from the word *constant*, because light travels at constant speed).

★ And many, many more men and women contributed to what we have today.

Learning never exhausts the mind.

— Leonardo da Vinci (1452–1519)

artist, musician, inventor, sculptor, engineer

??? Is Math Used after Graduation???

Math is used in everyday professions and living. Here are just a few examples.

Handling money in everyday transaction is extremely important.
For example when calculating:
1) change 2) profit 3) profit margin 4) pay on increases at a job 5) overtime at work 6) balancing accounts 7) inflation 8) purchase power (For example: What's a better deal? 2 hamburgers for $5.50 or 1 for $3.00) 9) discounts when purchasing 10) total wages after taxes 11) tax forms 12) Cell phone time used each month 12) video games scores. This list could be carried on for several pages.

Investing your money is another area of importance.
1) Calculating rate of return on dividends (a dividend of 10 cents on a $10 stock is smaller dividend than 5 cents on a $4 stock.
2) Calculating % change in prices (one mutual fund costing $50 per share goes up $2 and another mutual fund costing $20 goes up $1.50. Which is a better deal?
3) Calculating price to earning ratios.
4) Calculating cost basis.
5) Calculating compounding interest.
6) Amount obtained after 1 year on a CD at a certain percentage of interest.

Marine navigating uses math extensively.
1) Converting hours and minutes to decimals.
2) Determining time and distance calculations.
3) Calculating distance by measuring an angle to a known height.
4) Calculating distance by relative bearings.
5) Calculating vectors for example traveling north at 5 MPH for 2.5 hours with a south west current of 1.4 MPH.
6) Adding and subtracting seconds and minutes (example; 70 seconds = 1 minute and 10 seconds).
7) Converting true compass bearings to magnetic compass bearings.

Carpentry and construction is also a big user of math.
1) Adding fractions when measuring.
2) Calculating costs of materials used in construction.
3) Figuring costs especially when paying for a variety of lumber sizes by board feet.
4) Using the 3-4-5 triangle for building at right angles.
5) Converting degrees of slope to % of slope or to inches per foot or to millimeters per meter.
6) Converting fractions to decimals when measuring.
7) Calculating volume of complex shapes. For example; how much concrete is needed when to fill a form or shape?
8) Calculating rise and run of roofs.
9) Figuring how many steps are needed to get to a certain height.

Here are a few other professions and areas that use math.

✓Accountants
✓Agriculture
✓Architecture
✓Astronomy
✓Automotive industry
✓Banking industry
✓Business owners
✓Chefs
✓Chemistry
✓Clerks, cashiers
✓Clothing industry
✓Cooking
✓Computer programming
✓Department of Water and Power
✓Dentists
✓Engineers

✓Financial managers
✓Film industry
✓Food industry
✓Forensic scientists
✓Games (sports)
✓Gems and jewerly
✓Graphic designers
✓Insurance companies
✓Legal services
✓Librarians
✓Manufactoring
✓Math teachers
✓Medicine
✓Military
✓Music
✓NASA
✓Nurses

✓Plumbers
✓Optomitrist
✓Police and security
✓Printing
✓Public transportation
✓Real estate
✓Retail sales
✓Sales
✓Scientist
✓Sports
✓Video games
✓Waiter/waitress
✓Weatherman
✓Writer/author
✓And many many more jobs and professions

Testimonials

I was overhearing my son being taught the basics of math. I could not believe how simple it was. I never had a clue when I was in school about these basics. If I only knew back then what I learned in a few minutes, my education would have been less stressful.

—Parent of high school student

After two hours of tutoring, I was able to raise my algebra test scores from 30% to 93%.

—Francisco, 9th grade student

RJ made sure I understood math. I have passed my SAT scores and have been accepted to a couple of universities.

—Gerald, 12th grade student

I actually learned something that I never really thought of until I was asked. I thought they were very complicated words (math, arithmetic, algebra, geometry and calculus). *If you really take the time to understand what everything means, then math is so easy and that's what happened to me. I'm actually enjoying math right now and before I didn't like it.*

—Darinka, 12th grade student

Something amazing happened to me while working with math. A lot of people make it out to be hard stuff. Even I have considered math to be my worst subject in school. But when I really, really understand what I'm doing, especially the words, it's a wonder to me why I even hated math in the first place.

—Jordan, 12th grade student

I got what math is for the first time. I studied math for seven years and didn't know the things I needed to do well.

—Cameron, 7th grade student

Chapter VIII

Discover Where You Failed in Math

These tests are designed to find your strong points and weak points in arithmetic, algebra and geometry. It is suggested you recall which grade you were last doing well in math. For example, if 5th grade was when you remember math was fun and you were getting good grades, then start with the 5th grade math quiz. If you can't think of anytime math was easy and fun, then start with a lower grade and do all the tests from that point.

Somewhere along the line while doing these, you may have a hard time with certain problems. This area and the area prior to it should be taken up and understood well. Let us say on the 7th grade quiz you were unable to do a pre-algebra problem; then work on that area or check prior to it. These problems are designed to check your skill, so no calculators please.

A series of books will be published on the different areas of math. If algebra is a difficult area, then it is suggested you get the future publication, "Unlock the Mystery to Algebra."

These tests have been designed to take up the major areas of math. The tests do not constitute a comprehensive look at the whole subject of math. Textbooks with many different problems are available in libraries and on the Internet.

If you still have difficulties, then consult a teacher or professional tutor who can help you. A list of websites and professionals can also be found on page 95 of this book. The answers to the quizzes can be found on pages 86–91.

Good luck!

1st Grade Arithmetic

Circle the number <u>five</u>.   8   2   6   5   7

Circle the number <u>eight</u>.   1   5   7   8   2

<u>Add.</u>

0	1	7	3	8	5	4
+5	+2	+1	+3	+1	+2	+0

<u>Subtract.</u>

5	9	7	6	4	3	8
−4	−5	−7	−1	−0	−2	−6

<u>Counting.</u>

Which number is smallest?   8   6   7   9   3   4   5   0

Fill in the missing number.

A) 5, 6, __, 8   B) 1, __, 3   C) 8, __, 10

D) 3, 4, __, 6   E) __, 7, 8   F) 2, 3, 4, __

<u>Geometry</u>

Draw a line to each shape.

Square   Circle   Triangle   Rectangle

   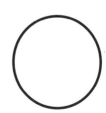

1st Grade Arithmetic (continued)

<u>Measuring.</u>

How many centimeters long?

A) 6 Cm   B) 5 Cm   C) 7 Cm   D) 8 Cm

How many inches long?

A) 5 Inches   B) 6 Inches   C) 3 Inches   D) 4 Inches

Jimmy's finger is 6_____ long.   Centimeters or Inches?

A fork is 7_____ long.   Centimeters or Inches?

<u>Problem Solving.</u>

Katie and her friends have 7 kites. They lost 2. How many kites were left?

_____

Gordon has 8 pennies. He found 5 pennies. How many pennies does he have?

_____

Sue has 8 apples. She ate 1. How many apples are left?

_____

Bret hit 9 golf balls. He lost 4 balls. How many golf balls does he have left?

_____

Michelle has 3 puppies. She gave one to a friend. How many puppies does she have now? _____

2nd Grade Arithmetic

Write in number form.

Fifty-eight _____   Nineteen _____   Forty-two _____   Sixty-three _____

Nine thousand, sixty-two_____Ten thousand, one hundred_____

Measuring.

How many inches in 2 feet?   _____

How many feet in 1 yard?   _____

Problem solving.

Rosanne has 2 coins. The total amount is 26 cents. What are the 2 coins?

_____

Colleen has 4 coins. The total amount is 61 cents. What are the 4 coins?

_____

There are 2 cups in 1 pint. How many cups in 2 pints? _____

There are 400 apples in a truck. Tommy took 200 apples from the truck. How many apples are still in the truck?   _____

Marge has 25 cents. She saved 25 more cents. How much does she have?
_____

Write the number.   2 tens 4 ones _____   8 tens   7 ones _____

Seven tens five ones _____   one ten   six ones _____

2nd Grade Arithmetic (continued)

Add.

10	29	80	39	27
+2	+31	+26	+11	+ 35

Subtract.

11	64	10	12	52
−8	−48	− 7	− 12	−23

Multiply.

9	4	10	3	1	8	7
x8	x3	x10	x2	x1	x6	x7

Fractions.

Shade one-half (½) of each shape.

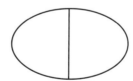
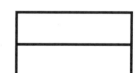

Shade one-fourth (¼) of each shape.

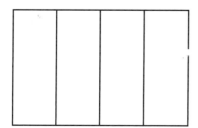
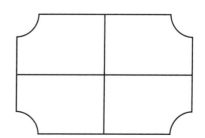

3rd Grade Arithmetic

Add.

```
  58        95        78         9        77        23        81
 +33       +56       + 6       + 68      +49       +65       + 11
```

Subtract.

```
  63        89        43        97        88        79        55
 -61       -43       - 0       -23       -44       -65       - 34
```

Multiply.

```
  12        79        10         7        51         4       348
 x 5       x 7       x 0       x 9       x 4       x 8       x 6
```

Put each decimal from least to greatest.

6.39   7.85   6.04   8.44   6.99

_____   _____   _____   _____   _____

Write in standard form.

Six hundred twenty-four thousand, ten   _____

Three thousand, twenty-four   _____

Fifty-six thousand, two hundred, ninety-one _____

3rd Grade Arithmetic (Continued)

Write as a decimal.

2 1/2 _____   3 1/10 _____   1 1/4 _____

Write as a fraction.

3.2 _____   2.5 _____   6.1 _____

Circle the fraction that shows the shaded part.

2/5   3/4   3/5   2/2   1/2   2/1

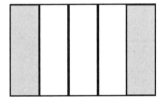   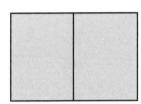

Show < or >   780 ____ 681   9,345 ____ 9,341   34 ____ 43
(< is less than)
(> is more than) 760 ____ 759   19,336 ____ 1,933   10 ____ 31

Divide

6)‾48‾   8)‾40‾   4)‾31‾   5)‾50‾

Solve

$6.25 $57.06 $10.95
- $1.10   + $ 4.21   + $ 1.04

Gordy wants a skateboard. It costs $58.95. He has $25.75 in his savings.
How much more money does he need to buy the skateboard? _____

4th Grade Arithmetic

<u>Name these shapes.</u> (Square, Rhombus, Rectangle, Parallelogram)

_____   _____   _____   _____

<u>Problem Solving</u>

✎ Karrie has $10. Can she buy 3 gifts that cost $3.88 each?

✎ A movie is 2 hours and 5 minutes long. Kellie is going to the 5:30 PM show. What time will the movie end?

✎ Paulie wants a hamburger, fries plus a drink. The hamburger is $2.65, fries are $1.85 and a soda is $1.65. If she has only $6.00, can she buy the 3 items?

✎ Jeanne solved a division problem. 183 ÷ 3 = 61
  Which one could Jeanne use to check her answer?

  A) 61 x 183   C) 183 x 3

  B) 3 x 61   D) None of the above

✎ Brian has a candy box measuring 12 inches by 6 inches by $\frac{1}{2}$ inch. What is the volume of the box? _____

4th Grade Arithmetic (continued)

Fractions.

4/6 + 1/6 = _____   2/5 + 1/5 = _____ $\frac{6}{8} = \frac{?}{4}$

4/9 + 5/9 = _____   1/3 + 1/3 = _____

1/10 + 7/10 = _____   1/2 + 1/4 = _____

Add.

```
     512          944,089          634          783,987
    + 46         + 45,867         + 74         +  6,434
```

Subtract.

```
   92,735           975          $9.76          694,340
 -  30,402         - 125         - $6.28        - 84,444
```

Multiply.

```
   $74.88           67            290            345
   x     6         x 22          x 13          x  53
```

Divide.

```
  32 ) 801        18 ) 900        31 )806
```

5th Grade Arithmetic

Solve.

$$\begin{array}{r} 334 \\ + 169 \\ \hline \end{array} \qquad \begin{array}{r} 649{,}861 \\ + 322{,}211 \\ \hline \end{array} \qquad \begin{array}{r} 876{,}431 \\ - 234{,}522 \\ \hline \end{array} \qquad \begin{array}{r} 74{,}799 \\ - 56{,}734 \\ \hline \end{array}$$

$$\begin{array}{r} \$29.30 \\ + \quad .79 \\ \hline \end{array} \qquad \begin{array}{r} 49.1 \\ + 5.8 \\ \hline \end{array} \qquad \begin{array}{r} 6.7 \\ - 4.3 \\ \hline \end{array} \qquad \begin{array}{r} 7.3 \\ - 4.2 \\ \hline \end{array}$$

$$4\overline{)70.8} \qquad\qquad 5\overline{)6.85} \qquad\qquad 9\overline{).810}$$

5/6 of 18 inches = _____   2/3 of 24 feet= _____

$$\begin{array}{r} 280 \\ \times\ 11 \\ \hline \end{array} \qquad \begin{array}{r} .19 \\ \times 88 \\ \hline \end{array} \qquad \begin{array}{r} 1\ 5 \\ \times\ 0.4 \\ \hline \end{array} \qquad \begin{array}{r} 2\ 8 \\ \times\ 0.5 \\ \hline \end{array}$$

What is the perimeter and area of these shapes? (Geometry)

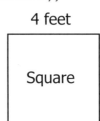

7 cm

2 cm

4 feet

Square

_____

_____

_____

_____

72

Change to mix fraction.

$\frac{26}{8}$ = $\frac{35}{10}$ = $\frac{82}{9}$ = $\frac{5}{4}$ =

Change to improper fraction.

6½ =   5¾ =   2¼ =

Multiply fractions.

3 1/2 X 3 3/4 =   1/5 X 3 1/4 X 5/6 =

Add, subtract fractions.

```
   4  2/3            6  2/8           10  3/6            7  7/10
 + 5  2/6          +5  1/4          −  7  1/12        −2  2/5
```

Write <,> or = in each.

0.23 _____ .25   3. 489 _____ 3.49   4.7000 _____ 4.7

5/8 _____ 3/12   6/22 _____ 3/11   2/8 _____ 7/32

Problem Solving.

Lea plans to drive from Madison to Chicago. It's a 2 hour drive. Gas costs $3.85 per gallon. From this information can she determine how much gas she will use? _____

Chris and Katie went fishing. They caught an average of 3 1/2 fish per hour. They caught a total of 14 fish. How many hours did they fish? _____

Banner has 3 construction sites to visit in one day. He works an 8 hour day. Traveling time between the sites is 45 minutes. If he visits each site for 2 hours will he have enough time to visit each site in 8 hours? (Note: He starts his 8 hour day at the first site) _____

6th Grade Arithmetic

Problem solving.

Ami wants to travel to Israel. His luggage weighs 60 pounds. The airline charges a fee of $25 plus $10 for each additional 10 pounds over 50 pounds. What will he pay for luggage charge?

    A) $75   B) $35   C) $60   D) $25

A new roller coaster ride "The Monster" carries 60 people on each ride. Steve and Ted are in line. There are 420 people are ahead of them. How many rides will it take before they get on?

    A) 10   B) 6   C) 8   D) 7

There is 296 miles of coast line along Oregon. The state of Washington has 157 miles and California has 840 miles of coast line. If Jim wants to travel the Western United States coast line how many miles will he go?

    A) 1,193 miles   B) 1,203 miles   C) 1,393 miles   D) None of these

Write these fractions as decimals.

    9/15   91/100   7/40   7/125

Write each pair of fractions using the least common denominator.

$$\frac{3}{4} , \frac{5}{8} \qquad\qquad \frac{5}{10} , \frac{11}{20} \qquad\qquad \frac{7}{8} , \frac{2}{20}$$

Solve.

    3/10 + 2/3 =   1/4 + 1/3 =   8 ½ + 2 ¾ =

    1/2 − 1/8 =   8 1/32 − 7 1/4 =   9 − 5/8 =

6th Grade Arithmetic (continued)

Write in standard form.

One hundred forty-six millionths.

A) .000146   B) .00000146   C) 146,000,000   D) .146

Three hundred four and twenty-seven hundredths.

A) 304.27   B) .3427   C) 304.027   D) 2,700.304

Solve.

$$\begin{array}{r} .68 \\ \times .40 \\ \hline \end{array}$$
$$\begin{array}{r} 4,095 \\ \times 245 \\ \hline \end{array}$$
$$\begin{array}{r} 0.0051 \\ \times 8.22 \\ \hline \end{array}$$
$$\begin{array}{r} 0.058 \\ \times .034 \\ \hline \end{array}$$

$5,133 \div 3 =$ _____ $6,700 \div 50 =$ _____ $148 \div .4 =$ _____

Gloria is making a dip for vegetables with ½ cup of cottage cheese and ½ cup of whole milk. How much calcium is in the dip?

A) 395   B) 198

C) 654   D) 327

	Amount	Calcium
Cottage cheese 1% fat	1 cup	138 ⸱ 69
Whole Milk	½ cup	258

6th Grade Arithmetic (continued)

Solve.

1/4 x 2/3 = _____   7/10 x 1/8 = _____   1/2 x 2 = _____

3 ÷ 1/4 = _____   4 1/4 ÷ 3/8 = _____   9 7/8 ÷ 2 1/3 = _____

4 pints = _____ quarts   24 quarts = ____ gallons

Word Problems.

Maya and her friends Lisa, Jessica and Alicia want to buy a Pizza. Maya wants 1/8, Jessica wants 2/8, Lisa wants 1/4 and Alicia wants 1/8 of the pizza. How much of the pizza is left?

   A) 1/2   B) 1/4   C) 1/8   D) None left over

An ant can lift 40 times its weight. If the strength of a person was of similar proportion, how much could Bill lift if he weighs 190 pounds?

A) 7,000 pounds   B) 7,600 pounds   C) 230 pounds   D) 70,000 pounds

Kay walked 5 miles in 2 hours. If she keeps the same pace, how far can she walk in 5 hours?

   A) 12 1/2 miles   B) 10 miles   C) 7 miles   D) 12 miles

Write the percent as a ratio.   10% _____   56% _____

Write the ratio as a percent.   3/5 _____   1/200_____

Lydia wants to buy a new purse at the mall. If the purse is $250, how much will she save if there is a 15% discount?

   A) $15   B) $37.50   C) $235.00   D) None of these

7th & 8th Grade Math

Complete.

6 meters = _____cm   8 liters = _____mL

9,000 grams = ____kg   18 inches = ____ft _____in

Multiply.

2,858	$2.89	0.0046	296.0
x 265	x 31	x .35	x 63

Divide.

471 ÷ .1000 = _____   12) 48.0096   5¼ ÷ 6 ⅞ = _____

Mandie has a babysitting job at her next door neighbor. She worked 4¾ hours and made a total $23.75. How much did she make per hour? _____

The surface of Mars temperature can vary 51 degrees centigrade. If the coldest is −84 degrees C, what is the warmest temperature. _____

Add/subtract.

¾ + (−½)= _____   (−¾) + (−⅝)= _____   − 1.3 + 7= _____

⅞− (−⅝)= _____   − 4.3 − (−3.3)= _____   ½ + ⅜ = _____

7th & 8th Grade Math (continued)

Solve

$$\frac{3}{5} = \frac{k}{15}$$

k = _____

$$\frac{0.54}{0.67} \overset{?}{=} \frac{1.05}{1.407}$$ _____

$$\frac{3}{8} \overset{?}{=} \frac{24}{64}$$ _____

20% of 8 is what?

A) 16 B) 1.6 C) .16 D) 2

Write as a decimal.   ½%

A) .5   B) .05 C) .005 D) .0005

Geometry

What is the formula for the area of a circle?

A) A = r²   B) A = π r   C) A = π r²   D) A = π r³

Find the perimeter and area of this triangle.

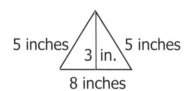

5 inches   3 in.   5 inches

8 inches

Perimeter_____   Area _____

Find the surface area of this cylinder to the nearest tenth. (S.A. = 2πr²+πdh)

4.5 feet

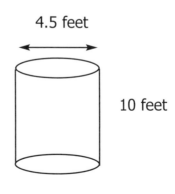

10 feet

Surface Area_____

Pre-algebra

Evaluate these expressions when c = 4 and d = 9.

$$\frac{1\ 7 + 2c - 2d}{7}$$

A) 1   B) 7   C) 0   D) 1½

$$\frac{10c + 4(-d)}{2}$$

A) 38   B) 2   C) 4   D) 1

Solve.

24n + 76 = 88   10 x −2 = 22   −3y −13 = 143

n = _____   x = _____   y = _____

$$\frac{a}{12} = -2$$

a = _____

Graph   x−y = 4

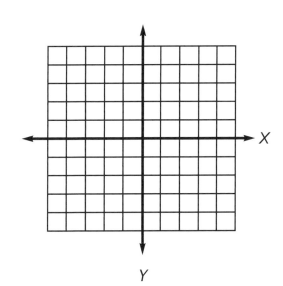

−14t = 60

t = _____

A business had a loss of $50 in November, a profit of $600 in December and a loss of $150 in January. What was the total profit for the 3 months.

_____

Algebra

<u>Solve.</u>   5p = −50   p = ?

    A) 10   B) −10   C) 1   D) None of these

$5 \frac{1}{3} = y + \frac{1}{4}$ y = ?

    A) $5 \frac{1}{2}$ B) $4 \frac{1}{2}$ C) $5 \frac{1}{12}$ D) $5 \frac{1}{1}$

−4.87 = x + .025   x = ?

    A) −4.62   B) −4.895   C) −5.12   D) 0

$3(6-4) \div 2 = ?$

    A) 12   B) 16   C) 3   D) 2

$\frac{63-3}{9+1} = ?$

    A) 0   B) no solution   C) 6   D) $1 \frac{1}{3}$

$25[2(3+7)] = ?$

    A) 500   B) 157   C) 325   D) 198

$6-(-5) = ?$

    A) 11   B) 1   C) −30   D) 30

$-9^2 = ?$

    A) −81   B) 81   C) 18   D) −18

$(-3)^3 = ?$

    A) 27   B) 9   C) −9   D) −27

$\frac{x}{2} = \frac{-4}{8}$ x = ?

    A) 1   B) −1   C) −4   D) −4

$\frac{-12}{20} = \frac{3}{y}$ y = ?

    A) −5   B) −4   C) −6   D) 6

Algebra (continued)

Solve. $\dfrac{\sqrt{4}}{\sqrt{25}}$

A) $\dfrac{2}{5}$ B) $\dfrac{2}{12.5}$ C) $\dfrac{2}{2.5}$ D) $\dfrac{2}{2}$

Complete. $5^2 \cdot 5^\square = 5^{11}$ $7^5 \cdot 7^2 = $ _____ $12^0 \cdot 12^1 = $ _____

Simplify. $(x^{-2})^2 (3xy^2)$

Solve. $w + 5 \leq 8$

A) $w < 3$ B) $W \leq 3$ C) $W > 3$ D) $w > 3$

$0 > k - 2.25$

A) $k > 2.25$ B) $k < 2.25$ C) $k < 0$ D) $k > 0$

Luther's Auto Repair bought special equipment for smog inspections. It cost is $1,200. The customer pays $60.00 for a smog check. The auto repair owner has a profit of $28.50 per inspection. How many inspections are needed to pay for the new equipment?

A) 40   B) 41   C) 43   D) 42

Bobby needs to do 17 hours of community service for a merit badge in Boy Scouts. He has completed 9 hours so far. What percentage has he done?
(Round to nearest thousandth)

A) 52.9%   B) 52.94%   C) 52.941%   D) 52.9411%

Algebra (continued)

Find the range when the domain is {2,0,4}.   g= 5p +2

Graph each function. y = 3 x+7   f(x) = x²

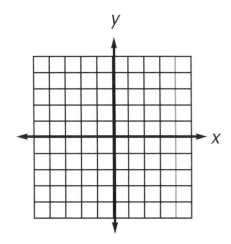

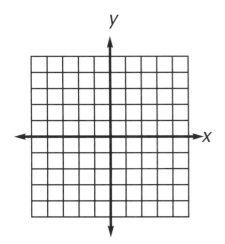

Mary has a summer job at a local computer store. A 5% commission is given each time she sells a computer. She sold a computer at $2,100. How much commission did he make?

   A) $210   B) $21   C) $105   D) $10.50

Tell whether the parabolas open upward or downwards.

   y= x² _____   y= −6x² _____   −y= x²+12 _____

Solve using the substitution method.

                y= x + 6.1
                y= −2x − 1.4

Solve.   √6.25

   A) 2.5   B) .625   C) 3.125   D)no solution

82

Geometry

Dave's swimming pool is 35.5 feet long by 25.5 feet wide by 72 inches deep. How many gallons are needed if the pool is filled to 50 inches from the bottom. (Note: 1 cubic foot is approximately 7.5 gallons)_____

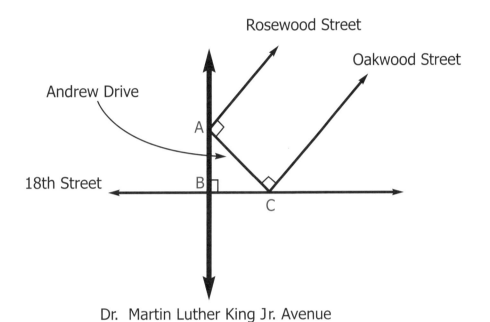

Dr. Martin Luther King Jr. Avenue

Which street is perpendicular to Dr. Martin Luther King Jr. Avenue? _____

Which roads are parallel to each other?   _____

Which roads form a 90 degree angle?   _____

If the road between A & B is 300 yards and the road between B & C is 400 yards long, how long is Andrew Drive? (A to C)   _____

If Rosewood and Oakwood kept running would they eventually intersect?_____

If Angle BCA is 40 degrees, how many degrees is angle BAC ?   _____

How many degrees is the angle formed by Dr. Martin Luther King Ave and Rosewood?   _____

A cylinder is 5 inches in diameter and 1 foot 4 inches high. What is the volume of the cylinder?

_____

Geometry (continued)

Find the slope if a line passes through

$(-5, 2)$ and $(-1, 4)$ $(0, 4)$ and $4, -2)$

A) 1   B) 2   C) 1/2   D) 1/4   A) 1/2   B) −1/2   C) −3/2   D) −2

A basketball is 9.78 inches in diameter. What is the surface area and volume of the basketball? Find to the nearest tenth. (S.A. $= 4\pi r^2$ V $= 4/3\ \pi\ r^3$)

_____   _____

Label each of the following equations. Does it form a straight line, circle, ellipse or a parabola?

A) $y = 2 + 6x^2$ _____

B) $y = -3x + 8$ _____

C) $x^2 + y^2 = 625$ _____

D) $\dfrac{x^2}{64} + \dfrac{y^2}{81} = 1$ _____

What is the volume of a cone that is 10 inches high? The diameter of the base is 6 inches. (Note: Volume of cone $= BH/3$)

A trapezoid is

A) a shape with the interior angles totalling 360 degrees.
B) a parallelogram.
C) a shape where all the angles are equal.
D) none of the above.

It's been found that Pi (π) is a repeating decimal. True or False?

Geometry (continued)

Find the perimeter and area of this parallelogram.

Perimeter _____
Area   _____

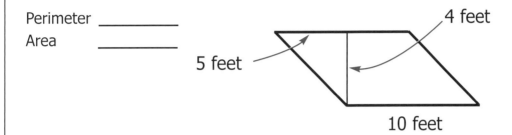

5 feet

4 feet

10 feet

Find the area of this table top excluding the hole (3 inch diameter) in the center. The Diameter of the table is 4.5 feet. _____

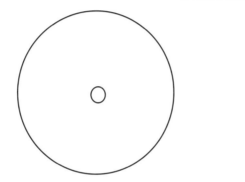

A six-sided polygon is called a _____.

   A) Hexagon   B) Rhombus   C) Pentagon   D) parallelogram

Which angle is Acute, Right and Obtuse?

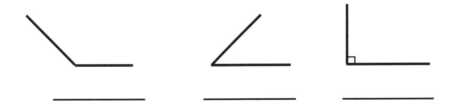

_____   _____   _____

Can a triangle have 3 angles that are each 65 degrees? _____

If two of the angles of a parallelogram are 60 and 120 degrees. What are the sizes of the other 2 angles? _____

Circle the number <u>five.</u>   8   2   6   5   7   **Page 64**

Circle the number <u>eight.</u>   1   5   7   8   2

<u>Add</u>

0	1	7	3	8	5	4
+5	+2	+1	+3	+1	+2	+0
5	3	8	6	9	7	4

<u>Subtract</u>

5	9	7	6	4	3	8
-4	-5	-7	-1	-0	-2	-6
1	4	0	5	4	1	2

Which number is smallest?   8   6   7   9   3   4   5   0

Fill in the missing number.

A) 5, 6, <u>7</u>, 8   B) 1, <u>2</u>, 3   C) 8, <u>9</u>, 10

D) 3, 4, <u>5</u>, 6   E) <u>6</u>, 7, 8   F) 2, 3, 4,<u>5</u>

<u>Geometry</u>

Draw a line to each shape.

Square   Circle   Triangle   Rectangle

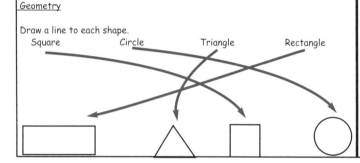

How many centimeters long?
A) 6 Cm   B) 5 Cm   C) 7 Cm   D) 8 Cm

How many inches long?
A) 5 Inches   B) 6 Inches   C) 3 Inches   D) 4 Inches

Jimmy's finger is 6_____ long.   <u>Centimeters</u> or Inches?

A fork is 7_____ long.   Centimeters or <u>Inches</u>?

<u>Problem Solving</u>

Katie and her friends have 7 kites. They lost 2. How many kites were left?

5 Kites

Gordon has 8 pennies. He found 5 more. How many pennies does he have?

13 Pennies

Sue has 8 apples. She ate 1. How many apples are left?

7 Apples

Bret hit 9 golf balls. He lost 4 balls. How many golf balls does he have left?

5 golf balls

Michelle has 3 puppies. She gave one to a friend. How many puppies does she have now? 2 puppies

Write in number form.

Fifty-eight <u>58</u>   Nineteen <u>19</u>   Forty-two <u>42</u>   Sixty-three <u>63</u>

Nine thousand, sixty two <u>9,062</u>   Ten thousand, one hundred <u>10,100</u>

How many inches in 2 feet?   <u>24 inches</u>

How many feet in 1 yard?   <u>3 feet</u>

Rosanne has 2 coins. The total amount is 26 cents. What are the 2 coins?

<u>1 quarter and 1 penny</u>

Colleen has 4 coins and the total amount is 61 cents. What are the 4 coins?

<u>2 quarters, 1 dime and 1 penny</u>

There are 2 cups in 1 pint. How many cups in 2 pints? <u>4 cups</u>

There are 400 apples in a truck. Tommy took 200 apples from the truck. How many apples are still in the truck? <u>200 apples</u>

Marge has 25 cents. She saved 25 more cents. How much does she have?
 <u>50 cents</u>

 Write the number.   2 tens 4 ones <u>24</u>   8 tens 7 ones <u>87</u>

              seven tens five ones <u>75</u>   one ten six ones <u>16</u>

 <u>Add</u>

10	29	80	39	27
+ 2	+31	+26	+ 11	+ 35
12	60	106	50	62

<u>Subtract</u>

11	64	10	12	52
-8	-48	- 7	- 12	- 23
3	16	3	0	29

<u>Multiply</u>

9	4	10	3	1	8	7
x8	x3	x10	x2	x1	x6	x7
72	12	100	6	1	48	49

<u>Fractions</u>

Shade one-half (½) of each shape.

Shade one-fourth (¼) of each shape.

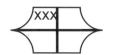

Add

58	95	78	9	77	23	81
+33	+56	+ 6	+ 68	+49	+65	+11
91	151	84	77	126	88	92

Subtract

63	89	43	97	88	79	55
-61	-43	- 0	-23	-44	-65	-34
2	46	43	74	44	14	21

Multiply

12	79	10	7	51	4	348
x 5	x 7	x0	x9	x4	x8	x6
60	553	0	63	204	32	2,088

Put each decimal from least to greatest.

6.39	7.85	6.04	8.44	6.99
6.04	6.39	6.99	7.85	8.44

Write in standard form.

Six hundred twenty four thousand, ten   624,010

Three thousand, twenty four   3,024

Fifty six thousand, two hundred, ninety one   56,291

Write as a decimal

2 1/2 2.5   3 1/10 3.1   1 1/4 1.25

Write as a fraction

3.2   3 1/5   2.5   2 1/2   6.1   6 1/10

Circle the fraction that shows the shaded part.

2/5   3/4   3/5   2/2   1/2   2/1

Show < or >   780 _>_ 681   9,345 _>_ 9,341   34 _<_ 43
(<, less than)
(>, more than)   760 _>_ 759   19,336 _>_ 1,933   10 _<_ 31

Divide

$\frac{8}{6)48}$	$\frac{5}{8)40}$	$\frac{7\ R\ 3}{4)31}$	$\frac{10}{5)50}$

Solve

$6.25	$57.06	$10.95
− $1.10	+ $ 4.21	+ $ 1.04
$5.15	$61.27	$11.99

Gordy wants a skateboard. It costs $58.95. He has $25.75 in is savings. How much more money does he need to buy the skateboard? $33.20

Name these shapes. (Square, Rhombus, Rectangle, Parallelogram)

Rectangle   Rhombus   Parallelogram   Square

Problem Solving

☐ Karrie has $10. Can she buy 3 gifts that cost $3.88 each? No

☐ A movie is 2 hours and 5 minutes. Kellie is going to the 5:30 PM

show. What time will the movie end? 7:35 PM

☐ Paulie wants a hamburger, fries plus a drink. The hamburger is

$2.65, fries are $1.85 and a soda is $1.65. If she has only

$6.00, can she buy the 3 items? No

☐ Jeanne solved a division problem: $183 \div 3 = 61$

Which one could Jeanne use to check her answer?

A) 61×183 C) 183×3

B) 3×61 D) None of the above

☐ Brian has a candy box measuring 12 inches by 6 inches by ½ inch. What

is it's volume? 36 cubic inches

Fractions

$4/6 + 1/6 = 5/6$ $2/5 + 1/5 = 3/5$ $\frac{6}{8} = \frac{3}{4}$

$4/9 + 5/9 = 1$ $1/3 + 1/3 = 2/3$

$1/10 = 7/10 = 4/5$ $1/2 + 1/4 = 3/4$

Add

512	944,089	634	783,987
+ 46	+ 45,867	+ 74	+ 6,434
558	989,956	708	790,421

Subtract

92,735	975	$9.76	694,340
− 30,402	− 125	− $6.28	− 84,444
62,333	850	$3.48	609,896

Multiply

$74.88	67	290	345
x 6	x 22	x 13	x 53
$ 449.28	1474	3770	18,285

Divide

$\frac{25\ R\ 1}{32) 801}$	$\frac{50}{18) 900}$	$\frac{26}{31) 806}$

Solve

334	649,861	876,431	74,799
+ 169	+ 322,211	−234,522	− 56,734
503	972,072	641,909	18,065

$29.30	49.1	6.7	7.3
+ .79	+ 5.8	−4.3	−4.2
$30.09	54.9	2.4	3.1

17.7	1.37	.09
4)70.8	5)6.85	9).810

5/6 of 18 inches = _15 inches_   2/3 of 24 feet= _16 feet_

280	.19	1 5	2 8
X 11	x 88	x 0.4	x 0.5
3,080	16.72	6.0	14.0

What is the perimeter and area of these shapes? (Geometry)

Perimeter 18 cm   Perimeter 16 feet

Area 14 square cm   Area 16 square feet

Change to mix fraction.

$$\frac{26}{8} = 3\frac{1}{4} \qquad \frac{35}{10} = 3\frac{1}{2} \qquad \frac{82}{9} = 9\frac{1}{9} \qquad \frac{5}{4} = 1\frac{1}{4}$$

Change to improper fraction.

$$6\frac{1}{2} = \frac{13}{2} \qquad 5\frac{3}{4} = \frac{23}{4} \qquad 2\frac{1}{4} = \frac{9}{4}$$

Multiply fractions.

3 1/2 X 3 3/4 = $\frac{105}{8}$ or 13 1/8   1/5 X 3 1/4 X 5/6 = $\frac{13}{24}$

Add, subtract fractions.

4 2/3	6 2/8	10 3/6	7 7/10
+ 5 2/6	+5 1/4	- 7 1/12	-2 2/5
10	11 1/2	3 5/12	5 3/10

Write <,> or = in each.

0.23 _<_ .25   3. 489 _<_ 3.49   4.7000 _=_ 4.7

5/8 _>_ 3/12   6/22 _=_ 3/11   2/8 _>_ 7/32

Problem Solving.

Lea plans to drive from Madison to Chicago. It's a 2 hour drive. Gas costs $3.85 per gallon. From this information can she determine how much gas she will use? _NO_, you would need to know the total miles and how many miles per gallon.

Chris and Katie went fishing. They caught an average of 3 1/2 fish per hour. All together they caught 14 fish. How many hours did they fish? 4 hours

Banner has 3 construction sites to visit in one day. He works an 8 hour day. Travel ling time between the sites is 45 minutes. If he visits each site for 2 hours will he have enough time to visit each site? Yes

Problem solving.

Ami wants to travel to Israel. His luggage weighs 60 pounds. The airline charges a fee of $25 plus $10 for each additional 10 pounds over 50 pounds. What will he pay for luggage charge?

A) $75   B) $35   C) $60   D) $25

A new roller coaster ride "The Monster" carries 60 people each ride. Steve and Ted are in line. There are 420 people are ahead of them. How many rides will it take before they get on?

A) 10   B) 6   C) 8   D) 7

There is 296 miles of coast line along Oregon. The state of Washington has 157 miles and California has 840 miles of coast line. If Jim wants to travel the Western United states coast line, how many miles will he go?

A) 1193 miles   B) 1203 miles   C) 1393 miles   D) None of these

Write these fractions as decimals.

9/15 .6   91/100 .91   7/40 .175   7/125 .056

Write each pair of fractions using least common denominator.

$\frac{3}{4}$, $\frac{5}{8}$ $\frac{6}{8}$, $\frac{5}{8}$	$\frac{5}{10}$, $\frac{11}{20}$ $\frac{10}{20}$, $\frac{11}{20}$	$\frac{7}{8}$, $\frac{2}{20}$ $\frac{35}{40}$, $\frac{4}{40}$	

Solve:

3/10 + 2/3 = 29/30   1/4 + 1/3 = 7/12   8 1/2 + 2 3/4 = 11 1/4

1/2 − 1/8 = 3/8   8 1/32 − 7 1/4 = 25/32   9 − 5/8 = 8 3/8

Write in standard form.

One hundred forty-six millionths.

A) .000146   B) .00000146   C) 146,000,000   D) .146

Three hundred four and twenty seven hundredths.

A) 304.27   B) .3427   C) 304.027   D) 2700.304

Solve:

.68	4095	0.0051	0.058
x.40	x 245	x 8.22	x .034
. 272	1,003,275	.041922	.001972

5,133 ÷ 3 = 1,711   6,700 ÷ 50 = 134   148 ÷ .4 = 370

Gloria is making a dip for vegetables with ½ cup cottage cheese and ½ cup of whole milk. How much calcium is in the dip?

A) 395   B) 198

C) 654   D) 327

Solve.

$1/4 \times 2/3 = 1/6$ $7/10 \times 1/8 = 7/80$ $1/2 \times 2 = 1$

$3 \div 1/4 = 12$ $4\,1/4 \div 3/8 = 11\,1/3$ $9\,7/8 \div 2\,1/3 = 4\,13/56$ or $237/56$

4 pints = 2 quarts   24 quarts = 6 gallons

Maya and her friends Lisa, Jessica and Alicia want to buy a Pizza. Maya wants 1/8, Jessica wants 2/8, Lisa wants 1/4 and Alicia wants 1/8 of the pizza. How much of the pizza is left?

A) 1/2   B) 1/4   C) 1/8   D) None left over

An ant can lift 40 times its weight. If human strength were a similar proportion, how much could Bill lift if he weighs 190 pounds?

A) 7000 pounds   B) 7600 pounds   C) 230 pounds   D) 70,000 pounds

Kay walked 5 miles in 2 hours. If she keeps the same pace, how far can she walk in 5 hours?

A) 12 1/2 miles   B) 10 miles   C) 7 miles   D) 12 miles

Write the percent as a ratio. 10% 1/10   56% 14/25

Write the ratio as a percent.   3/5 60%   1/200 .5%

Lydia wants to get a new purse at the mall. If the purse is $250, how much will she save if there is a 15% discount?

A) $15   B) $37.50   C) $235.00   D) None of these

Complete.

6 meters = 600 cm   8 liters = 8000 mL

9000 grams = 9 kg   18 inches = 1 foot 6 inches

Multiply.

2858	$2.89	0.0046	296.0
x 265	x 31	x .35	x 63
757,370	$89.59	.00161	18,648

Divide.

$471 \div .1000 = 4{,}710$ $12 \overline{)48.0096} = 4.0008$ $5\,1/4 \div 6\,7/8 = 42/55$

Mandie has a job babysitting for her next door neighbor. She worked 4¾ hours and made a total $23.75. How much did she make per hour? $5.00 per hour

The surface of Mars temperature can vary 51 degrees Centigrade. If the coldest is −84 degrees C, what is the warmest temperature. −33

Add/ subtract.

$3/4 + (-1/2) = 1/4$ $(-3/4) + (-5/8) = -1\,3/8$ $-1.3 + 7 = 5.7$

$7/8 - (-5/8) = 1\,1/2$ $-4.3 - (-3.3) = -1$ $1/2 + 3/8 = 7/8$

Solve

          NO   YES

$\dfrac{3}{5} = \dfrac{9}{15}$ $\dfrac{0.54}{0.67} = \dfrac{1.05}{1.407}$ $\dfrac{3}{8} = \dfrac{24}{64}$

20% of 8 is what?   Write as a decimal. ½%

A) 16 B) 1.6 C) .16 D) 2   A) .5 B) .05 C) .005 D) .0005

Geometry

What is the formula for the area of a circle?

A) $A = r^2$ B) $A = \pi r$ C) $A = \pi r^2$ D) $A = \pi r^3$

Find the perimeter and area of this triangle.

      Perimeter 18 inches   Area 12 square inches

Find the surface area of this cylinder to the nearest tenth.

Surface Area 173.1 square feet

Pre-algebra

Evaluate these expressions when c = 4 and d = 9.

$$\dfrac{17 + 2c - 2d}{7}$$ $$\dfrac{10c + 4(-d)}{2}$$

A) 1 B) 7 C) 0 D) 1½   A) 38 B) 2 C) 4 D) 1

Solve.

$24n + 76 = 88$ $10x - 2 = 22$ $-3x - 13 = 143$

n = 1/2   X = 2.4   X = −52

$\dfrac{a}{12} = -2$

a = −24

$-14t = 60$

t = −4 2/7

Graph x - y = 4

A business had a loss of $50 in November, a profit of $600 in December and a loss of $150 in January. What was the total profit for the 3 months. $400.00 profit

Algebra

Solve. 5p = −50

A) 10   B) −10   C) 1   D) No answer

5 1/3 = y + 1/4

A) 5 1/2   B) 4 1/2   C) 5 1/12   D) 5 1/1

− 4.87 = x + .025

A) −4.62   B) −4.895   C) −5.12   D) 0

3(6−4)÷2 = ?

A) 12   B) 16   C) 3   D) 2

$\frac{63-3}{9+1}$ = ?

A) 0   B) no solution   C) 6   D) 1 1/3

25[2(3+7)]= ?

A) 500   B) 157   C) 325   D) 198

6−(−5)= ?

A) 11   B) 1   C) −30   D) 30

−9² = ?

A) −18   B) 81   C) 18   D) −81

(−3)³ = ?

A) 27   B) 9   C) −9   D) −27

$\frac{x}{2} = \frac{-4}{8}$

A) 1   B) −1   C) −4   D) −4

$\frac{-12}{20} = \frac{3}{y}$

A) −5   B) −4   C) −6   D) 5

Algebra (continued)

Solve. $\frac{\sqrt{4}}{\sqrt{25}}$

A) $\frac{2}{5}$ B) $\frac{2}{12.5}$ C) $\frac{2}{2.5}$ D) $\frac{2}{2}$

Complete. $5^2 \cdot 5^9 = 5^{11}$ $7^5 \cdot 7^2 = 7^7$ $12^0 \cdot 12^1 = 12$

Simplify. $(x^{-2})^2 (3xy^2) = \frac{3y^2}{x^3}$

Solve.   w + 5 ≤ 8

A) W < 3   B) W ≤ 3   C) W > 3   D) w > 3

Solve.   0 > k − 2.25

A) k > 2.25   B) k < 2.25   C) k < 0   D) k > 0

Luther's Auto Repair bought special equipment for smog inspection. It cost $1200. The customer pays $60.00 for an inspection. The owner of the garage gets to keep $28.50 on each inspection. How many inspections are needed to pay for the equipment?

A) 40   B) 41   C) 43   D) 42

Bobby needs to do 17 hours of community service for a merit badge in Boy Scouts. He has completed 9 hours so far. What percentage has he done?
(Round to nearest thousandth)

A) 52.900%   B) 52.940%   C) 52.941%   D) 52.9411%

Algebra (continued)

Find the range when the domain is {2,0,4}. g= 5p +2 {2,12} {0,2} {4,22}

Graph each function. y=3x+7   f(x)=x²

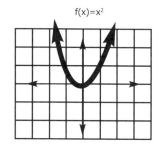

Mary has a summer job at a local computer store. A 5% commission is given each time a computer is sold. She sold a computer for $2100. How much commission did she make?

A) $210   B) $21   C) $105   D) $10.50

Tell whether the parabolas open upward or downwards.

y=x² UP   y=−6x² DOWN   −y=x²+12 DOWN

Solve using the substitution method.

y=x+6.1   Y = 3.6
y=−2x−1.4   X = -2.5

Solve: √6.25

A)3.125   B) .625   C) 2.5   D)no solution

Geometry

Dave's swimming pool is 35.5 feet long by 25.5 feet wide by 72 inches deep. How many gallons are needed if the water is filled to 50 inches from the bottom of the pool? (Note: 1 cubic foot is approximately 7.5 gallons) 28,289.1 gallons

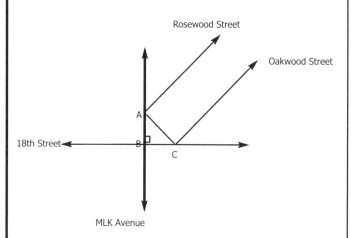

Which street is perpendicular to MLK Avenue? 18th Street
Which roads are parallel to each other?   Rosewood and Oakwood
Which roads form a 90 degree angle? 18th & MLK , Andrew Drive & Oakwood, Andrew Drive & Rosewood
If road A & B is 300 yards and B & C is 400 yards long, how long is the road from A to C? 500 yards
If Rosewood and Oakwood kept running would they eventually intersect? NO
If angle BCA is 40 degrees, how many degrees is angle BAC? 50 degrees
How many degrees is formed by the angle of MLK Ave and Rosewood? 40 degrees

A cylinder is 5 inches in diameter and 1 foot 4 inches high. What is the volume of the cylinder? 314 cubic inches

Find the slope if a line passes through:

     (-5,2) and (-1,4)   (0,4) and 4,-2)

 A) 1   B) 2   C) 1/2   D) 1/4   A) 1/2   B) -1/2   C) -3/2   D) -2

A basketball is 9.78 inches in Diameter. What is the surface area and volume of the basketball?
(Note: S.A.. = 4 π r^2, V = 4/3 π r^3)
Surface area is 330.3 square inches.   Volume is 489.4 cubic inches

Label each of the following equations whether it forms a straight line, circle, ellipse or a parabola.

   A) y = 2 +x^2 Parabola

   B) y = -3x+8   Straight line

   C) x^2 + y^2 = 625   Circle

   D) $\dfrac{x^2}{64} + \dfrac{y^2}{81} = 1$ Ellipse

What is the volume of a cone that has is 10 meters slant height and has a base of 6 meters diameter? (Note: Volume of cone is area of base times height divided by 3) 89.86 cubic meters

A trapezoid is:
A) a shape with the interior angles totalling 360 degrees.
B) a parallelogram.
C) a shape where all the angles are equal.
D) none of the above.

It's been found that Pi (π) is a repeating decimal. True or False?   False

Find the perimeter and area of this parallelogram to the nearest tenth.

  Perimeter is 32.8 feet
  Area is 40 square feet

Find the area of this table top to the nearest hundredth. The diameter is 4.5 feet. Excluding the hole in the center which is a

3 inch diameter. 15.85 Square feet

A six-sided polygon is called a?

    A) Hexagon   B) Rhombus   C) Pentagon   D) parallelogram

Which angle is Acute, Right and Obtuse?

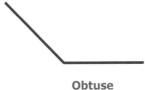

                 Obtuse   **Acute**   **Right**

Can a triangle have 3 angles that are each 65 degrees? NO

If two of the angles of a parallelogram are 60 and 120 degrees. What are the sizes of the other 2 angles? 60 and 120 degrees

GLOSSARY

Algebra: A branch of math using numbers and letters to solve problems. It also means reunion or completion.

Arabic numerals: Numbers we use throughout the world today.
(0, 1, 2, 3, 4, 5, 6, 7, 8, 9)

Arabia: A peninsula in the Middle East that is bordered by the Red Sea, Persian Gulf and Arabian Sea.

Area: The measurement of surface of an enclosed shape. For example, the area of a rectangle is calculated done by multiplying the two adjacent sides.

Arithmetic: A branch of math which includes adding, subtracting, multiplying and dividing.

Calculus: A branch of math that is a combination of arithmetic, algebra and geometry, used to solve complicated problems.

Coefficient: A number before a letter used in algebra and calculus. For example, $3x$. 3 is the coefficient.

Decimals: A fraction with the denominator being 10, 100 etc.: it is shown by a point before the numerator (example 5/10 is .5).

Diameter: The distance of a straight line through the center from one side of a circle to the other.

Equations: Numbers and letters used in all branches of math with an equal sign.

Exponent: The number of times a value is multiplied. For example: 3^2 is 3 x 3. 2 is the exponent.

Expression: A combination of letters and numbers to show a value. For example, (3 + X) is an expression.

Factoring: It means multiplying. If you factor 3(X+4) it would be 3X+12.

GLOSSARY

Fractions: A part of a whole or group. Fractions are shown as one number over another, for example 1/2.

Geometry: A branch of math dealing with space. It means the measure of Earth.

Inequalities: *In-* means "not" and *equality* means "equal," so it means *"NOT EQUAL."* Symbols used are < (less than) and > (greater than).

Hindu numerals: Numbers (0, 1, 2, 3, 4, 5, 6, 7, 8, 9) developed around AD600 in India.

Math: Using numbers and letters to solve problems dealing with distances, amounts, time, energy and space.

Percentages: *Per-* means *for every. Cent* means *one hundred. For every hundred.* Example: 35% would be 35 for every one hundred.

Perimeter: The total distance around the shape. Adding all the sides together will give the peremeter of the shape.

Radius: The distance of a straight line from the center to the edge of a circle.

Simplify: Reduce an expression to put into simplest terms. For example 2X + 3X can be stated as 5X.

Square root: The reverse of exponents. If 3 squared is 9, then the square root of 9 is 3. It can also be written this way: $\sqrt{9} = 3$

Will: A statement that lays out the distribution of property of a deceased person.

About the Author

RJ **Toftness** has been an educator/tutor for over 30 years. He received his bachelor of science degree in education from the University of Wisconsin—River Falls.

After home schooling his children and others, plus tutoring and instructing over a thousand students in various subjects throughout the years, he came to realize that a basic step was missing in math. After extensive research into the subject he located a fundamental step had been removed from math in the 1960s. Teaching these basics, he found students gaining a greater understanding and appreciation for mathematics.

From an early age he sought to learn more about the practical aspects of living. With much help from his parents he achieved the Eagle rank in Boy Scouts. From volunteering in Habitat for Humanity projects to his travels in India to educate young adults, gave him a wider scope on the importance of a practical education. He once stated, "An Education is as valuable as you can achieve your dreams."

Communicate to the author? Visit the website at www.math-unlock.com or e-mail rjpublishers@yahoo.com.

Recommended Books, Websites and Tutors

Books

Key to Fractions and Algebra
by Julie King & Peter Rasmussen

Heath Mathematics
by Rucker/Dilley/Lowry

Heron Books of Math
P.O. Box 503
Sheridan, Or 97378

Beginning Algebra with Application
by Linda Exley & Vincent Hall

Arithmetic Refresh
by A.A. Klaf 1964 edition

WEBSITES

www.tutor.com

www.math.com

PBS.org

www.dr.math.com

Tutors and Professionals

Success! Tutoring of Burbank California
info@success-tutoring.com

Sylvan Tutoring
www.sylvan-learning-center.com

Tutors
www.appliedscholastics.org

Unlock the Mystery to Math

<u>**Order Form**</u>

☎ Telephone Orders: Call   (323) 665-6657

💻 Internet Orders: www.math-unlock.com

✉ Postal Orders: RJ Toftness
554 N. Mariposa Ave.
Los Angles, CA. 90004

Fill out the following:

Name: _____

Address:_____ Apt # _____

City:_____ State:_____Zip:_____

Telephone:_____ E-mail:_____

Check or money order (no COD) U.S. currency only payable to: **RJ Publishing**
Charge: (check one) ☐ VISA ☐ Master Card   ☐ Discover

Card number   __ __ __ __   __ __ __ __   __ __ __ __   __ __ __ __

Expires_____/_____   _____
       month   year   Card holder signature

Sales tax: (Only California residents, add 8.25% = $1.56).   _____

Shipping: $4.00 for first book and $2.00 for each additional book. _____

Book price _____ $18.95 _____

Total cost: _____

Visit www.math-unlock.com for information on future publications titled:
Unlock the Mystery to Algebra & *Unlock the Mystery to Fractions*